CAMRA'S
HOME-BREWING
PROBLEM SOLVER

Quarto is the authority on a wide range of topics.

Quarto educates, entertains and enriches the lives of our readers—enthusiasts and lovers of hands-on living.

www.QuartoKnows.com

First published in the United Kingdom in 2017 by the Campaign for Real Ale Ltd
230 Hatfield Rd
St Albans
Herts
AL1 4LW
www.camra.org.uk/books

10 9 8 7 6 5 4 3 2 1

ISBN: 978-1-85249-347-9

Produced by
QUID Publishing Ltd
Part of the Quarto Group
1st Floor, Ovest House
58 West Street, Brighton
BN1 2RA
www.quidpublishing.com

Designed by Clare Barber
Written by Erik Lars Myers
UK home-brewing consultant: Andy Parker

Printed in China

CAMRA'S
HOME-BREWING
PROBLEM SOLVER

Erik Lars Myers

CAMRA
BOOKS

CONTENTS

INTRODUCTION

Home-brewing can be an incredibly rewarding hobby that is a fantastic blend of both art and science. It is, however, not without its trouble points.

This book covers some of the most frequent problems that you might run in to while making beer at home, whether you're using an extract kit or a complex all-grain mash. The intrepid brewer will encounter many more problems that aren't contained specifically here, but the shadows of these are almost definitely to be found among these pages.

You'll find out what to do when ingredients have gone bad, when equipment isn't quite working the way it's supposed to, or

how to deal with yeast that isn't behaving quite right. You'll find information about how to diagnose problems after your beer is finished and packaged, and – crucially – how to avoid those problems in the future. In some cases, you may want to change your process, in others your expectations. Sometimes you might discover that it's time to upgrade some equipment.

You may use this book in whatever way suits you and your needs. Whether you pick a problem and open the book right to that page, or read the book from front to end as a guide of what (or what not) to do, it will contain the basics of making great beer at home, as well as some helpful tips and tricks to make your life easier as you do, alongside in-depth information about styles and even the science of beer.

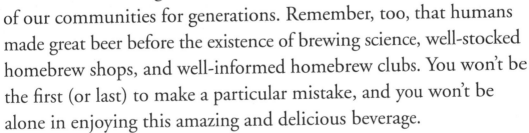

Finally, remember that humans have been making beer for thousands of years. Brewing is a human tradition that has been with our species from the dawn of agriculture, present at both celebrations and tragedies. It's been at the centre of our communities for generations. Remember, too, that humans made great beer before the existence of brewing science, well-stocked homebrew shops, and well-informed homebrew clubs. You won't be the first (or last) to make a particular mistake, and you won't be alone in enjoying this amazing and delicious beverage.

CHAPTER ONE
THE BASICS

While the actual concept behind making beer is fairly simple – add water to barley to create and extract sugar, add hops while boiling, cool then feed to yeast – the actual execution can seem complex and confusing. There can be a lot of equipment involved, storage needs and specialised knowledge required. For many hobbyists, homebrew can seem like a return to biology, chemistry, arithmetic and physics long left behind. It can seem intimidating.

A good brewer is a scientist: you must follow formulae consistently, measure frequently, and keep good and accurate records of your methods and results.

A great brewer is also an artist: using the flavour and nuance of a particular malt or a hop at a specific time, in the same, delicate way that a painter might apply brushstrokes across a canvas.

In both the art and science of brewing, a careful, methodical approach is necessary, whether it's for achieving a perfect mash temperature, or knowing the correct measurement required to make a beer just right. Take the time to learn, and do it right every time.

Finally, in brewing, cleanliness is king. It's often said that brewing is 90 per cent cleaning and 10 per cent actually making beer. Every great beer starts with clean and sanitised equipment.

01 I don't know how to start home-brewing

CAUSE

Making beer is a blend of both art and science. With the wide variety of ingredients, equipment and literature available it can be an intimidating task to start the hobby.

SOLUTION

In recent years home-brewing has become a thriving hobby industry, which means you will find more technology and resources available than ever before. To simplify things, there are just a few things you need to get going with homebrew:

- Brewing equipment: a pot to boil in, a vessel to ferment in, a kitchen scale, and a thermometer (more details in Problem 2).
- Packaging equipment: casks, kegs or bottles and bottlecaps (and a capper).
- Ingredients: barley, hops, water and yeast (starter kits are available from homebrew shops and will contain everything you need to get started, including detailed instructions).

While there are hundreds of books available, many containing valuable information, seeking advice from others in a social setting can be the easiest way to start. Find a local homebrew club, and ask questions. There you'll find beginners intermediates, and experienced veterans who can answer questions about getting started, allow you to sit in on a brew day with them and share some of their recipes (and their results!). They can give you advice on what literature was most helpful for them, the best places to get ingredients locally, and how many bottles to bring to the next meeting.

Beer is a social beverage, and so it's only fitting that communities are often the best place to learn how to make great beer.

RELAX, DON'T WORRY, HAVE A HOMEBREW

The first widely known book written about home-brewing was *The Joy of Brewing*, by Charlie Papazian, first published in 1976. In it, among information on equipment, ingredients, sanitation, cleaning and fermentation, Charlie dispensed the most valuable advice on home-brewing: that when all else fails, you should relax; don't worry, have a homebrew.

Home-brewing is a process which can be simple or complicated. It's up to the individual brewer to decide what kind of set-up best suits their space and budget. No matter what, there will always be beer at the end.

02 I'm not sure if I have the right equipment

CAUSE

As home-brewing has become a popular hobby, the amount of equipment available to use at home has grown. For a beginner, it can be difficult to tell what equipment is necessary.

SOLUTION

A hobbyist can set up quickly, easily and cheaply with a simple set of tools:

- *A kettle*: this is a pot to boil the liquid in. If you intend to boil the full volume of wort that you create you will need a kettle that is slightly larger than your batch size. Try to avoid aluminium or copper pots, or non-stick coating. Stainless steel is best.
- *A kitchen scale:* for weighing ingredients, such as grain or hops. Check your scale measures are in the same units your recipe calls for (otherwise find a conversion chart or phone app).
- *A measuring cup:* a large measuring cup with small increments is most useful for recording precise water measurements.
- *A thermometer:* something that can measure from room temperature through boiling. Homebrew supply shops will have a selection of thermometers, but any calibrated meat or kitchen thermometer may do.
- *A spoon or stirring device:* something you can mix and stir with. Spoons or ladles that are made of metal or wood are best for avoiding burnt fingers and sticky clean-ups.
- *A fermentation vessel:* Whether a high-priced cylindroconical fermenter, or a plastic bucket, it must hold at least the volume of your fermentation with 10 to 25 per cent head space. It should be non-porous (plastic, glass or metal), able to withstand wide-swing temperature changes and robust cleaning chemicals, and have a means of venting excess CO_2. Bear in mind that scratches or dents in plastic vessels can be difficult to clean.

🌿 *Pictured here is some of the basic equipment needed for home-brewing: a fermentation bucket, a bubble airlock and hydrometer. There is a wide range of brewing equipment available for use at home. It is vital that everything is kept clean and stored well when not in use.*

STORING EQUIPMENT

Home-brewing can be an equipment-heavy hobby, which means that storing equipment can be a challenge. When possible, store equipment dry, after it has been cleaned, then sanitise it before the next use. If equipment is stored in an area with any moisture – particularly a closed environment – it can be an invitation for mould to grow or metal parts to rust.

03 I'm not sure which ingredients to buy

CAUSE

There are millions of potential ingredient combinations available to make beer. Finding the ones that make the beer that you want can be confusing.

SOLUTION

The easiest way to get started with ingredients is to use kits. A good homebrew kit can help familiarise you with how a recipe for a particular style is built, what ingredients typically go into each batch of beer, and in what proportion and order. Your local homebrew shop will carry an array of kits, and will sometimes be able to build a kit for a particular style on demand.

Kits are a good way to discover the individual ingredients that make up a beer. Once you've achieved good results from a particular kit, you can learn through experimentation. Try using the IPA kit you loved last time with different hops, or the same stout kit with a different type of dark grain – or with no dark grain at all. Experimentation in brewing – and drinking – is the best way to learn about the ingredients available, including how ingredients vary according to manufacturer.

Another good tactic is to find a 'clone recipe'. These are intended to mimic a commercial beer (and are often released by commercial brewers themselves). Assuming you use a faithful clone recipe (or at a least good one), creating clones can be an excellent method to learn how commercial brewers use ingredients, and are also a good means of measuring your brewing chops.

Taste, smell and feel all the ingredients that go into your kettle. As with cooking, sampling each of your ingredients can help you craft a better final product – or know the moment that something is going wrong.

KNOWING THE FLAVOUR
OF YOUR INGREDIENTS

Spend time looking up the manufacturer or source of each of your ingredients. Maltsters, hop farmers and yeast labs tend to publish flavour profiles for each of the products that they sell so that you can make the most educated purchase. As you use each ingredient, don't be afraid to taste, smell and touch them so that you know the experience of each ingredient.

It is beneficial for a brewer to be well versed in the component ingredients that goes into a beer, as well as the finished product.

04 I don't know how to build my own recipe

CAUSE

Moving on from homebrew kits to making your own recipes can be intimidating; there seems to be an endless variety of ingredients and variables to consider.

SOLUTION

To build a new recipe consider the following:

- Advanced brewers might go by flavour, but it can be easy to start from a style.
- In most cases, it's a good idea to use a pale, neutral malt or extract for most of the recipe, and then add character with speciality malts.
- Munich and Vienna malts add a rich malty, grainy character. Caramel or crystal malts make rich caramel and toffee flavours. Highly kilned malts add nutty flavours. Roasted malts can add coffee, chocolate or dark ashy flavours. Speciality malts should only be added in small proportions for nuance and subtlety.
- A low temperature (64–67°C/147–152°F) will make a highly fermentable, thinner-bodied beer. A high temperature (68–70°C/154–158°F) will make a full-bodied beer with fewer fermentable sugars. Or you can balance in the middle range.
- Small additions of high-alpha hops at the beginning of the boil can add lots of bitterness. Save hops for the end of the boil for aroma and flavour – or leave them out altogether.
- There are various strains of commercial yeast, each with their own flavour profile.
- Yeasts have their own ideal temperature ranges. At the bottom of that range, yeast will create a cleaner fermentation; at the top, more robust flavour compounds.

05 I can't tell if my fermentation is complete

CAUSE

Fermentation typically takes place in a closed vessel that shouldn't be disturbed or contaminated. It's sometimes difficult to know when fermentation is complete without causing an infection in your beer.

SOLUTION

The easiest way to tell if fermentation is complete is to watch the bubbles in the airlock of the fermenter. When yeast stops generating CO_2, it is typically at or near the end of fermentation but, to be sure, wait a day or two after the bubbles stop.

If you ferment in a clear-sided demijohn, you can easily view yeast action inside. During fermentation, yeast will be in solution and moving around. When fermentation is complete, yeast will settle to the bottom of the fermenter, and the liquid will be almost completely clear.

If you can, take a sample of beer from the fermenter, use a refractometer or a hydrometer/saccharometer to take a gravity reading. You will need to have measured the density of the beer prior to fermentation, and know the typical attenuation of the yeast strain you're using (i.e. how much of the available sugar yeast will eat). This information can be found printed on the package of yeast or through the manufacturer's website. If a yeast strain typically eats 65 to 75 per cent of the available sugar, you can use maths to find out how much of the sugar is left in your fermentation, and how close you are to target attenuation, using the formula: (Original Gravity – Current Gravity) ÷ Original Gravity.

Finally – taste it. If fermentation is complete, it should taste like beer, minus any carbonation. If there is an overt sugary sweetness or the beer still tastes like wort, fermentation has not been completed.

06 I don't know how much alcohol is in my beer

CAUSE

Individual recipes often list how much alcohol is expected in a beer, but knowing the precise amount is almost impossible on a homebrew scale.

SOLUTION

True measurements of alcohol content can only be done in a lab setting with expensive equipment. You can't know how much alcohol is in your finished beer without taking gravity, or density, readings (See Problem 5). Once you know the original gravity of your beer and the final gravity of your beer, you can make some very educated guesses.

The easiest way to calculate alcohol is to find a calculator on the Internet. There are hundreds of websites available that allow a homebrewer to insert original and final gravity figures into an online form, and return a result on the alcohol content. But how do you know if it's accurate?

There are many different methods to calculate alcohol content using an algorithm, and they can be quite complex. For easy home use, keep this handy formula around:

Alcohol by Volume = (Original Gravity - Final Gravity) × 131

For example: 1.048 − 1.011 = 0.037 × 131 = 4.8 per cent

To calculate 'alcohol by weight'

multiply the 'alcohol by volume' by 0.79336.

4.8% × 0.79336 = 3.8 per cent ABW

These calculations will be fairly accurate on mid-strength beers, but they will overestimate low-alcohol beers and underestimate high-alcohol beers. For more accurate results, find calculators that use either the Balling method or the DeClerk method.

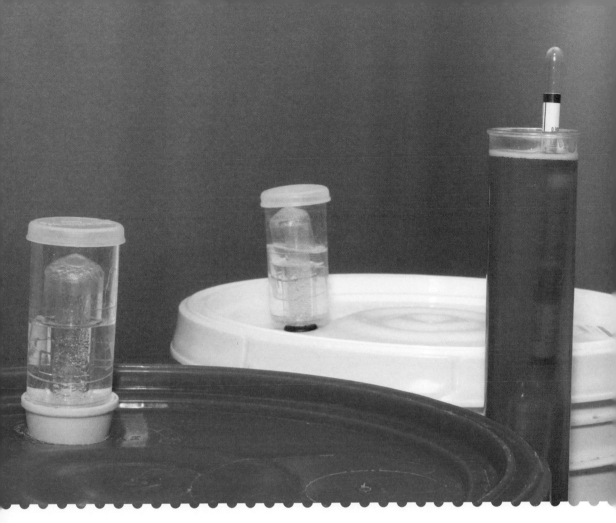

Monitoring airlock activity and taking hydrometer readings help to keep track of the fermentation progress.

SPECIFIC GRAVITY

Home-brewers commonly measure density in specific gravity using a hydrometer or saccharometer: a weighted glass cylinder which floats differently in liquids of varying density. The specific gravity of water at 15.5°C (60°F) is 1.000, which increases with the addition of sugar. For instance, a beer that is around 5 per cent alcohol by volume (3.9 per cent alcohol by weight), would usually start around 1.050, assuming an average attenuation of roughly 80 per cent. When doing calculations with specific gravity, it's often easier to drop the '1.0' from the specific gravity number and use the last two as your 'gravity units'. 1.050 would be 50 gravity units.

07 I need to take a sample out of my fermenter

CAUSE

Occasionally, a brewer may need to take a sample of the beer in a fermenter, but doing so may infect the beer with wild yeast or bacteria, if it is not handled correctly.

SOLUTION

Drawing samples from a fermenter can be useful to monitor the taste or to take a gravity reading, or to see if fermentation is complete. Some lucky home-brewers have fermentation vessels with spigots on the side that allow them to draw samples without exposing the fermentation to oxygen or to foreign objects, but most brewers do not have that luxury.

Samples are most often drawn with a device known as a 'thief' (or a 'wine thief' from the corresponding device in the wine industry): a glass cylinder, open on both ends, which is easy to clean and sanitise. To operate, submerge the bottom part of the thief into the fermenting liquid, allow it to fill, then place your thumb over the opening at the top and remove the thief, with beer inside.

In theory this is possible with any hose or cylinder, but the wider the diameter of the hole in the end of the thief, the more liquid will spill out when removed from the larger body of liquid. Wine thiefs are built to have a wide, cylindrical body with tiny openings at either end.

Alternatively, you could use a clean and sanitised glass measuring cup to remove a sample from the fermenter, or pour liquid out of the fermenter into another vessel. If pouring, be mindful of sanitising all surfaces that beer will move across, and also that this technique will likely stir up anything that has settled to the bottom of the fermenter as well as oxygenate the wort.

If it is necessary for a brewer to take a sample from the fermenter it is most important to take care and not contaminate the rest of the fermenting liquid. If using a hydrometer to measure fermenting beer, remember that there can be some carbonation present in the liquid. Spin the hydrometer to knock off any bubbles and get a more accurate reading.

08 I'm unsure how to clean my equipment

CAUSE

Home-brewing can be a messy business. Traditional cleaning chemicals and soaps may leave residues behind that can cause unpleasant flavours, disrupt head retention, or inhibit yeast growth in future batches.

SOLUTION

Cleaning homebrew equipment is of vital importance. Leaving equipment dirty can allow mould and bacteria to grow, leading to smelly homebrew storage and infected beer. The first thing to consider when cleaning your brewing equipment is what the potential infection risks are, as these will be the spots that require the most careful cleaning.

When cleaning equipment it's sometimes helpful to break it down into groups: hot side and cold side. Hot-side equipment only touches the beer as heat is applied, so is of lower risk to infection than cold-side – equipment which touches the beer after it has been cooled. It is vital to clean all equipment diligently, but pay close attention to cold-side equipment.

At the end of a brew day, be sure to rinse and remove any residual sugar or leftover bits of grain or hops from the inside and outside of all equipment. This can be accomplished outside the house with a hose or even in a clean bathroom. When scrubbing surfaces to remove debris, only use sponges or soft, natural scrubbers. Harsh scrubbing material like steel wool can scratch the surface of plastic objects, creating spaces for bacteria to grow.

If possible, invest in a non-foaming oxidative alkali cleaning chemical, which can be found at your local homebrew shop. This helps remove organics or any stains from your equipment and also makes cleaning faster and easier. Be sure to wear proper personal protective equipment, such as eye goggles and gloves, when handling cleaning chemicals, as these can be harsh on your skin, damaging soft tissues easily.

Cleaning is a vital part of brewing. Trub, yeast, and residual hops coagulate during fermentation.

ORDER OF OPERATIONS

It's much easier to clean equipment while it's still wet, or just after use, rather than much later when sugar or bits of grain have dried on, but it's not necessary to sanitise your equipment right away. Sanitised equipment is ready to use immediately. If you sanitise equipment and then let it dry for a week until your next brew day, it is no longer sanitised and will require re-sanitising prior to use.

09 Should I clean or sanitise my equipment?

CAUSE

Cleaning a piece of equipment is not the same as sanitising it, but many home-brewers are easily confused by the two actions because they appear to be very similar.

SOLUTION

You can't sanitise dirt. In the context of home-brewing, when the term 'cleaning' is used it refers to the removal of debris from homebrew equipment, which generally means getting rid of solid waste: sugar, grain, hop debris, or other bits left over from the brewing process. Once everything is cleaned, the brewer is left with two possible contamination vectors: air and water. Though modern water sources are clean and bioform-free, there remains the small possibility that some waterborne bacteria are present. Sanitisation is meant to cut down on the risk of infections caused by mould spores and airborne bacteria that can infiltrate an open vessel.

The equipment that needs to be sanitised is generally the equipment that will touch the beer after it has been chilled for fermentation. That means your fermenter, any hoses or transfer devices, or even a spoon or funnel needs to be sanitised prior to use. Hot-side equipment does not need to be sanitised because when temperatures reach boiling, fewer organisms can survive. After wort has been boiling for 10 minutes, it is effectively sterile.

Sanitisation is best done with chemicals that are specifically made for that purpose. If you rinse sanitiser off a surface, that surface is no longer sanitised. Once you finish sanitising a fermentation vessel, be sure it is closed until its next use so that nothing can contaminate it in the meantime. Items that are to be used in the open air (funnels, spoons, hoses, etc.) should be sanitised immediately prior to use.

SANITISED VS. STERILISED

A sanitised surface is clean and hygienic. A sterilised surface is completely free of bacteria or any other living organisms. Sterilisation is not likely to be achieved in a home environment. A healthy fermentation yeast will outcompete other organisms and create a hostile environment for most bacteria. There are no common human pathogens that can survive in beer, but to make the best-tasting beer, keep sanitisation up.

A clean demijohn is important but ensure it is also sanitised before adding beer to it.

CHAPTER TWO
INGREDIENTS

Beer is an agricultural product. The base of beer – the essences of barley malt and hops dissolved into water and fed to yeast – is an elegant, natural culmination of the Earth and its environment.

Keeping those ingredients fresh and using them well are among the first challenges that home-brewers face as they begin to brew. What starts off as a little extra grain left over from the weekend's brew is the beginning of a collection of disparate bags and boxes filled with a few ounces of speciality malt, scattered here or there, never going away, and never ultimately getting used in a recipe until weeks, months or even years later. A small bag of hops in the freezer evolves into a freezer drawer stacked with bits of pellets and hop dusts, waiting to be thrown into the annual 'kitchen sink' beer to make space for food in the freezer.

How each of these ingredients is handled and processed can make a crucial difference. Beer is, in so many ways, more than just the sum of its parts. It can amplify, exemplify, or heighten flavour, regardless of whether that flavour is great to begin with.

10 My hops smell like cheese or feet

CAUSE

Under the wrong conditions, the alpha acids in hops degrade and form isovaleric acid, which contributes a flavour and aroma reminiscent of Parmesan cheese, or sweaty socks, that can linger into the finished beer.

SOLUTION

Hops are very delicate. They can change flavour character and lose bitterness if not stored properly. Hops are best stored very cold (0°C /32°F or below) in an airtight container away from light. Both whole-cone and pelletised hops are sold in opaque, vacuum-sealed bags to lengthen their shelf life as much as possible. Even in the original unopened bag, hops should be stored in the freezer, as bittering compounds can decay quickly.

Once the bag is opened, you should try to keep the hops away from long-term air contact, as oxygen promotes the formation of isovaleric acid, particularly in warm conditions. Transferring unused hops to a resealable plastic bag which has had the air squeezed out of it as much as possible – or has been vacuum-sealed – will ensure the longest possible hop freshness. Stackable airtight plastic containers that can be placed in a freezer are perfect for hop storage. Be sure to label containers with date, variety and alpha-acid information for later reference.

Unfortunately, once a beer has been made with oxidised, or cheesy, hops the flavour will be in the beer, and there is no good way to remove it. Fortunately, it is not easily detectable in small amounts.

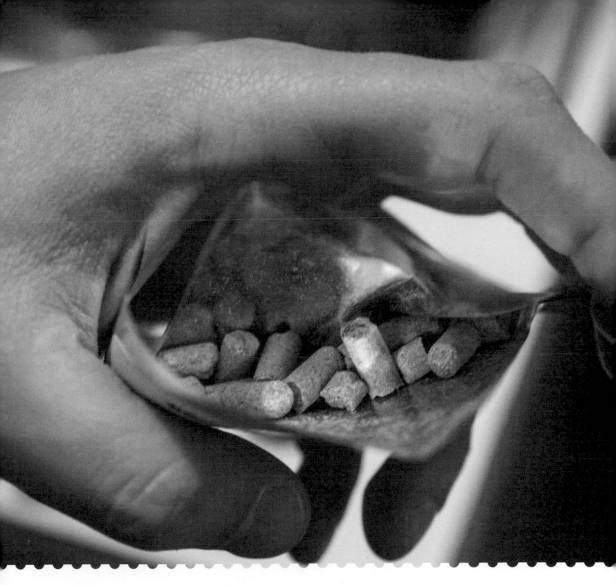

WHOLE CONE VS. PELLETISED HOPS

Whole cone hops are often used by home-brewers, as well as some of the largest craft brewers. They are thought to have much more nuanced aromas when used in dry hopping, and contribute fewer grassy flavours than pelletised hops. However, pelletised hops can be much easier to store in a small kitchen freezer they allow much more surface-area contact within the boil and are much easier to clean up at the end of a brew day.

🌺 *Hops can be easily stored in the bag they were purchased in. Simply seal the bag and keep it in the freezer.*

11 My malt extract has solidified

CAUSE

If dry malt extract is stored in moist conditions, or liquid malt extract is stored for a very long time, it tends to crystallise or solidify, which can make brewing difficult.

SOLUTION

The good news is that the malt extract is still perfectly good to use. The bad news is that the brew day just got longer.

Crystallised dry malt extract is much easier to deal with than crystallised liquid malt extract. Soak the extract in hot water (but not over an active heat) while occasionally stirring as the crystals break down. Once the crystallised malt has completely dissolved into the water, you can move on with the rest of the brewing.

When liquid malt extract solidifies, it's much more difficult to deal with, but still usable. Heat the malt extract in its original container using a double-boiler, which keeps it in a liquid bath. Don't use direct heat, which might scorch or burn the extract.

In both cases, measuring out a portion – if your recipe calls for it – is possible, but much more difficult, and it is ultimately easier to use the whole thing.

When malt extract solidifies or crystallises, some of the sugar molecules in the extract can change shape and become less fermentable. There is a chance that a beer made with old or poorly stored extract will be less attenuative than fresh, but in many or most cases it will be fine. However, if the extract smells or tastes mouldy or off do not use it to make beer; at best it will ruin the flavour, at worst it will infect your whole batch.

If using liquid extract in a recipe, heating the extract slightly before use can help it to flow more freely.

12 My malt extract is mouldy

CAUSE

If malt extract has been poorly stored and exposed to open air, mould spores may land in the nutrient-rich environment and grow, making the malt extract unsuitable for brewing.

SOLUTION

The easiest solution for this problem is to only buy enough malt extract to use in the batch of beer that you are currently working on, and always to use fresh extract. However, it's not always practical. Different recipes call for varying amounts of extract that may not match the increments the shop sells, meaning that a brewer may have some extract left over.

Once opened, liquid malt extract should be stored in a cool, dry place in an airtight container, and used as soon as possible. Many liquid malt extract containers are resealable, which makes storing an unused amount easier, but packages that are not resealable may require repackaging into a clean and sanitised container. Liquid malt extract is easier to transfer from one container to another when it is warm.

Dry malt extract can be stored in cool, dry storage, and in most cases can be stored in the container that it was purchased in – often a bag or a jar. If the container has become wet in any way, it is important to transfer the malt extract to a new cleaned, sanitised and thoroughly dry container. Sealable plastic kitchen containers are excellent for storing dry malt extract. Unless it has been exposed to moisture, dry malt extract has an excellent shelf life.

Liquid malt extract has a shelf life of about three months in storage; dry malt extract can keep for at least one year.

🌾 *Mouldy extract should never be used for brewing, but if it is properly stored, malt extract should not become mouldy anyway.*

13 My grain has bugs in it

CAUSE

If brewing grain has been stored poorly for a long period of time, or contaminated somewhere along the distribution chain, it can become infested by grain weevil or meal moth eggs and may no longer be suitable to use.

SOLUTION

The first sign of a grain weevil infestation is an excess of powdery husk pieces within grain, where the weevils have broken through the husk to eat. The weevils themselves are often easy to spot. They look like small ants at first glance but reveal themselves upon closer inspection. Meal moths can be trickier, as the moths tend to be very mobile. However, like other moths, the larvae do form a chrysalis within grain storage and may be detectable by small silk filaments on the side of a bag of grain.

Upon finding bugs in grain, all steps should be taken to stop the infestation so that it does not become an ongoing problem. Isolate or destroy the grain that contains the infestation, then thoroughly clean the area that the grain was stored in. Be sure to check and clean any nooks or cracks in porous surfaces, in case there are bugs hiding that will re-infest your grain.

Barley malt is a perishable agricultural product, and should be treated as such. Grain should be stored uncrushed in a sealed container in a cool, dry place to avoid infestation. Cool temperatures and lack of moisture can inhibit hatching cycles. Properly stored grain can last a year or more with no issues.

Grain with the occasional bug is okay to brew with, but large infestations can mean a loss of sugar and carbohydrates in the malt, as well as an elevated moisture content from the waste products of the insects. Elevated moisture content can lead to mould and considerable off-flavours.

Meal moth larvae can sometimes be difficult to spot among barley kernels, and if present can create long-standing problems in a home brewery.

14 My grain is mouldy

CAUSE

If brewing grain has been stored in a moist environment it can become infested with mould, which is inconvenient and wasteful.

SOLUTION

Under no circumstance should you brew with mouldy grain. Moist grain does not crush well, which means that brewing efficiency will suffer, and there is a significant chance for mould-related, musty or earthy off-flavours in your final brew. Some mould spores can survive boiling temperatures, which means that your final product may be infected immediately.

Many home-brewers only buy as much grain as they need to brew with, thereby removing the trouble of storing grain at home. However, buying in small quantities is not always possible, and buying in large quantities can often come with a bulk discount.

Grain should be stored in a cool, dry, dark environment. Large, opaque, resealable plastic containers are often the best place to store grain. Be sure that any bags that grain might be in are also tightly sealed within the containers. Whenever possible, keep different types of grain in separate containers.

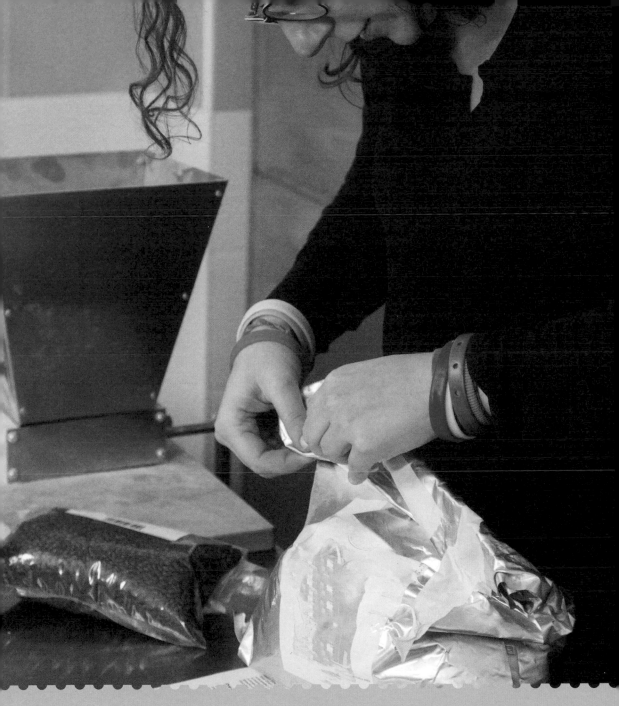

It is best to take precautions to avoid mouldy grain in the first place. Once opened, properly storing grain in air-tight containers away from sunlight can mean reliable long-term use and excellent beer in the future.

15 My grain won't crush well

CAUSE

The spacing of the rollers in the grain mill may be set too wide to properly crush grain, or the grain may be stale, or contain a higher moisture content, causing the husk of the barley to bend, instead of crack.

SOLUTION

If grain is not crushed well, it can be difficult to access the starch and proteins within, leading to low efficiency, and less sugar in the wort than intended. To diagnose whether the problem is the roller spacing or the moisture content of the grain, start by crushing some grain in a bowl or with a mortar and pestle, or in your mouth. Malted barley should be easy to crack or bite; it should not be chewy, gummy or bendable.

If the issue is high moisture content, check your grain is being stored correctly – in resealable, airtight, plastic containers in a cool, dry place. The moisture content of grain can be reduced by placing the grain in a warm oven, set to the very lowest temperature for short amounts of time – 15 to 20 minutes. Take special care to monitor the drying process, as oven-roasting grain can also toast the grains. While this can sometimes add a pleasant nutty or caramel character to malt, it can also burn malt very quickly. Toasted malt will be darker than anticipated when used in brewing.

If moisture content isn't the problem, it is time to check your mill. The gap in a homebrew roller mill should be set to 0.9 mm (0.0036 inches) and can be measured with a feeler gauge. Well-crushed grain kernels should be cracked open, revealing the starch – which looks like white dust – inside the kernel without fully crushing the husk itself into fine pieces. Husks should be largely intact.

If grain is not properly crushed it may well end up being less efficient or useless for brewing beer with.

16 I crushed my grain too finely

CAUSE

The roller gap on the barley mill was set too close together, crushing grain into a fine powder. Grain crushed too finely can lead to lautering problems, as well as possible off-flavours in the beer. Barley husks crushed too finely can lead to a harsh astringency.

SOLUTION

Pre-crushed grains purchased from homebrew shops should be the correct consistency for most applications. Many home-brewers prefer this approach to grinding their own grain. If you have ground the grain yourself, the first thing to check is the rollers on your mill. Rollers should be set 0.9 mm (0.04 inches) apart from each other, or about 0.9mm as measured by a feeler gauge. If a feeler gauge is unavailable, a credit card can suffice.

Finely crushed grain means a slow lauter because it lacks the barley husks needed to create channels within the grain bed that allow water to pass through. You must now create that matrix of fibre to allow water to pass through. To address lautering problems, mix rice hulls or uncrushed barley into your grain – or your mash bed – as much as possible.

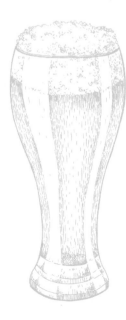

Rice hulls are very light and can be difficult to mix into a mash bed after the mash is wet, but they are inexpensive and add no character to a finished beer, so large amounts may be used to help lautering.

Uncrushed barley malt can often help build a good mash bed because the kernels can act as whole husks. It can also help bolster efficiency by a very small amount. If the uncrushed barley is not base malt, some additional flavour or colour may be contributed to the beer. The cost of using uncrushed barley is considerably higher than rice hulls, and it may take much more uncrushed barley than it would rice hulls because they are not as long.

WHAT ARE RICE HULLS?

Rice hulls are the husks from the outside of a rice grain. They do not add sugar, colour or flavour to a mash, so can be an great addition to high-adjunct or high-protein beers to help brewing run more smoothly. Normal usage rate is 10 per cent of the volume of unhusked grain, though many brewers brew with much smaller percentages. Use trial and error to find the perfect ratio.

Ideally grain should feature whole, uncrushed husks as well as finely crushed white dust. This dust is the starch extracted from the kernel by the crushing process.

17 My water smells funny or has a discolouration

CAUSE

Various water sources can cause unpleasant smells or colours, whether it's spring run-off in a well or rust-coloured water from an old iron pipe. Water with high mineral or biological content can effect pH, solubility and water hardness.

SOLUTION

It is best to brew with clean, pH-neutral (around 7 pH), good-tasting, soft water. If your normal water source – your tap water – is often discoloured or has an off-putting aroma or flavour, it is worth considering purchasing bottled water to brew with. If you trust the water to wash your dishes in, then the same water can be used to clean your brewing equipment. However, be extra-vigilant with sanitation prior to use. It takes only a tiny amount of bateria to cause contamination. As noted in Problem 9, never rinse a sanitiser – particularly in this case.

A more permanent solution might be a simple charcoal filter, installed either on your water tap or earlier in your plumbing system, to help remove water impurities.

In either case, the wisest course of action is to have your water tested for mineral content, heavy metal content and biological content. Simple water-test kits are available from most hardware shops, but often only indicate safety thresholds, rather than an exact make-up of your water. The most accurate water tests will require sending a sample of water to a lab for testing. It can take a few days to a few weeks to receive results, but this will allow you to make the most informed decisions about your water content. Water reports are often available through your local council.

HARD WATER VS. SOFT WATER

Hard water contains a large amount of dissolved mineral –
usually measured in parts per million (ppm). In brewing, hard
water often accentuates hop bitterness and can lead to an
apparent dryness in beer. Soft water has less than 100 ppm of
dissolved minerals, and is often preferred by brewers because
they can selectively add the most common water-soluble
minerals – carbonate, sodium, chloride, sulphate, calcium and
magnesium – to create a desired level of hardness depending
on brewing style.

Tap water is often fine to brew with. When in doubt, however, bottled water is a good alternative. Installing a carbon filter on a tap can also help with any water problems.

18 My water is chlorinated

CAUSE

Some municipalities use chlorine or chloramine to kill harmful coliform and bacteria, which can be metabolised by yeast to create chlorophenols, flavour compounds that can be detected by humans as 'antiseptic' or 'medicinal' flavours at as little as 5 parts per billion (ppm).

SOLUTION

If you live in an area with a public water system, you likely have some chlorine content in your water. It's up to you to assess your finished beer and make the decision on whether chlorine or chloramine removal is in your best interest.

Chlorine is volatile and easy to remove from water. Boiling water will allow chlorine to evaporate, and even stirring liquid in a pot or allowing water to sit out overnight can remove active chlorine content from it.

Chloramine, however, is much more stable, which is often why it is used in public water supplies. Chloramine stays in the water, killing bacteria all the way to your tap, and all the way to your brew kettle. Chloramine can be removed with a high-quality charcoal filter that allows long contact times as well as reverse-osmosis systems.

For quicker water treatments, brewers can use Campden tablets, which are usually potassium or sodium metabisulfites. The sodium metabisulfite will react with chlorine and chloramines and remove them from the water. About half to one tablet should be sufficient to dechlorinate about 23 litres of water.

CAMPDEN TABLETS

Campden tablets are named after the town Chipping Campden in Gloucestershire, England, where the compound was developed as a means of fruit preservation. It was originally a powder that was meant to be added to water to create a preservation solution. It was developed into a tablet format during the Second World War as an easy means of food preservation; one tablet was to be dissolved in a half-pint of water to preserve one pound of fruit.

Boiling water can easily reduce chlorine content, but chloramines will require extra help to get rid of.

19 The expiration date on my yeast has passed

CAUSE

Retailers didn't rotate expired stock, or a brewer neglected to use the ingredients when they bought it, leading to expired product. Once yeast has passed its printed expiration date, it may no longer be suitable for brewing.

SOLUTION

Unless the yeast in question has gone through extreme temperatures or conditions, or the expiration date passed years ago, this yeast will probably still be alive and able to make beer, although it may need some help.

There are two important factors in determining whether yeast is suitable for brewing: viability (how many of the cells in each sample are alive) and vitality (how healthy the yeast cells are). In a lab setting, a brewer could isolate a sample from these, place them on a hemocytometer with a dye, and perform a cell count to find out both viability and vitality. At home, it's a bit less exact.

If possible, it's safest to either replace this yeast with a fresh pitch or supplement it with an additional yeast pitch of the same type. Where more or replacement yeast is unavailable, it's important to make a yeast starter (see Problem 20). The yeast should be growing, off-gassing or creating krausen before it is pitched into finished beer. If there is no yeast activity in the starter, you may be increasing infection risk.

If starting with dry yeast, rehydrate it for 10 minutes in distilled water before creating a starter with it. When yeast is dried and dormant, it loses the ability to regulate what passes through its cell walls, so adding it directly to wort may significantly reduce viability. In an expired yeast pitch, that may include the remaining viable cells.

DRY YEAST VS. LIQUID YEAST

There are pros and cons to both dry and liquid yeasts. Dry yeast has a much longer shelf life, and there are many more cells available in a dry yeast pack than in a liquid vial or smack pack. However, liquid yeast vendors often have a much wider variety of strains available, giving the brewer a greater choice of yeast character to complement any recipe.

Liquid yeast can come in a wide variety of strains. The use of 'smack packs' of yeast (like the one pictured here) may seem like a good shortcut to creating yeast starters, but some additional help may be required to get the starter working.

20 I don't know how to make a yeast starter

CAUSE

Between stir plates, Erlenmeyer flasks and fancy stoppers, yeast starters can seem intimidating, but they can mean the difference between good beer and great beer.

SOLUTION

Making a yeast starter is as simple as putting a small amount of yeast into a portion of unfermented wort and allowing it to ferment. The primary purpose is to give the yeast time to grow more yeast and become active and healthy before being added into a large, sugar-rich environment.

Start with a cleaned and sanitised glass container that is large enough that your starter takes up two-thirds to three-quarters of the available space. The glass container needs a way to be covered so that debris, including bacteria or mould spores, do not fall in, but it needs to have a way to allow gas to escape. This can be a stopper with an airlock on it, a foam stopper, or even plastic wrap stretched across the opening of the jar and held loosely on with a rubber band.

Ideally, make sure that your yeast starter is between 1.040 and 1.050 specific gravity. It's good to start with sterile wort. Otherwise, boil a litre of water with 100 g of dried malt extract and a few hop pellets, then cool to room temperature. Once the yeast has been pitched into the starter, agitate the starter occasionally to keep yeast in solution. Remember to keep sterile everything the yeast will be in contact with.

Within a day, krausen should form on the top of the starter, and a layer of yeast should be visible on the bottom. Once the starter has finished its full fermentation, it is ready to pitch into your homebrew. It should be good for two or three days after fermentation has completed. You can store it in the refrigerator, but the starter should be the same temperature as the wort it will be going into when you pitch to avoid shocking the yeast cells.

PLANNING AHEAD FOR STARTERS

If you can, consider saving a portion of your wort as a starter for your next batch. Clean and sanitise a 1 litre (1 quart) sized canning jar. Fill the jar halfway with sterile wort, seal and save it in the refrigerator for your next yeast pitch. Just take care that you don't use a dark wort to make a starter for a light beer, because it will affect the colour of the next beer.

A good yeast starter can be made on a magnetic stir plate in an Erlenmeyer flask, or made in a jar on your kitchen counter — the basics remain the same.

21 I'm not sure how much yeast to put into my wort

CAUSE

Counting yeast cells at home can be challenging to impossible, so many home-brewers must take on faith that they pitched the correct amount of viable, healthy yeast.

SOLUTION

Over- or underpitching yeast can lead to fermentation problems, and undesirable flavours in the finished beer. A good rule of thumb for home-brewers is to pitch 1 million cells per millilitre of wort per degree Plato for ales, or 1.5 million cells per millilitre for lagers. This assumes that around half of the cells pitched will not be viable. Most home-brewer's yeasts promise 100 billion cells or more, and many offer viability upward of 90 per cent. One package of healthy yeast can ferment up to 5 gallons of wort (20 litres) at 10 plato.

Use this information to determine whether you have an ideal pitch size for your beer. The ideal operating window for yeast pitches is 500,000 to 2 million yeast cells per litre. If you are on the low end of this range, more yeast growth will occur before resources are consumed, leading to more yeast character in the final beer. If you are on the high end of the range, less growth will occur, meaning less yeast character in the finished beer (a 'cleaner' beer). Underpitching can lead to sluggish or incomplete fermentation, or high ester or fusel alcohol content. Overpitching can mean that fermentation completes before the yeast can clean up after itself, leading to lingering fermentation-related off-flavours.

While re-pitching from one homebrew to another can save money, it's best to start with a commercial yeast package with your final plato in mind, to be sure you have the correct amount of yeast in your pitch.

PLATO VS. SPECIFIC GRAVITY

Plato (P) is a measure of density that can be converted to specific gravity (SG). The formula to convert P to SG is fairly easy (SG = 1+ (Plato ÷ (258.6 − ((Plato ÷ 258.2 × 227.1))), and many conversion charts are available online. However, you can do a rough conversion by multiplying degrees plato by 4. So, 10 P would equal 1.040 SG. It's not always perfectly accurate, but it's close.

A hydrometer should always be read from the top of the meniscus of the liquid; i.e. where the liquid rises to meet the glass of the hydrometer.

CHAPTER THREE
MASHING AND STEEPING

Whether mashing an all-grain batch of beer, or steeping a bag of grain in preparation for an extract batch, the home brewer goes through the same process as the professionals: using heat to activate enzymes in water – a substrate that the grains work well in. These enzymes then break down long, complex hydrocarbon chains – that we know as starch – into small, bite-sized (for yeast) molecules: glucose, maltose and maltotriose. This process utilises the wonderful, dynamic range of colour, flavour and texture available in barley malt.

Problems here usually occur when temperatures aren't being met or maintained, when barley isn't crushed just right, or simply when the brewer has a forgetful moment. Most issues that a brewer runs into at this point in the process are easy to recover from. You'll still have beer at the end, but may have a much longer afternoon in front of the brew kettle than you originally planned. Having some homebrew handy on one of these afternoons can of course make them pass by much more quickly!

22 The alpha-acid content in my hops is different

CAUSE

While the same variety of hops will generally have a similar range of alpha-acid content, different growing conditions can cause a variance in bittering potential. Alternately, a brewer may want to substitute different hops into a recipe while maintaining the same bitterness character.

SOLUTION

The trick to making this calculation is to understand alpha acids. The alpha acid listed on the package of hops is the percentage of the dry weight of the hop that is alpha acids. Alpha acids undergo a chemical reaction during the boil, in which humulone is rearranged into trans- and cis-isohumulone, both of which are perceived by humans as bitter. The higher the alpha-acid content of the hop is, the fewer hops you need to create the same amount of bitterness.

On a home-brewing scale, the easiest way to make this calculation is to multiply the alpha-acid content of the hop listed in the recipe by the number of grams or ounces it calls for to find the Alpha Acid Units (AAU), and then divide the AAU by the alpha-acid content of the new hop.

For example, if your recipe calls for 100 g (3.5 oz) of East Kent Goldings (5.5 per cent AA), but you would like to substitute Bramling Cross (4.2 per cent AA), calculate the following:

$$3.5 \times 5.5 = 19.25 \text{ AAU}$$
$$19.25 \div 4.2 = 4.6$$

It will take 4.6 oz (130g) of Bramling Cross (4.2 per cent AA) to create the same amount of bitterness as 3.5 oz of East Kent Goldings (5.5 per cent AA).

🌿 *Whole-leaf hops, such as these, can be stored safely in an air-tight container. The bittering potential of your hops is dependent on the growing conditions of this natural product.*

WHAT'S AN IBU?

IBUs are a measure of the ppm of dissolved isomerised alpha acids in beer. 1 IBU is 1 ppm, or 1 milligram of iso-alpha acid per litre of beer. It is not an effective measurement of how bitter a beer will taste, but it is an excellent way to make consistent beer by ensuring the same recipe, made twice, contains the same number of IBUs.

23 I forgot to add my extract/all of my extract

CAUSE

Brew days can be busy and distractions can easily pop up, meaning even the most careful brewer can forget a step now and then.

SOLUTION

Even though this seems like an egregious error – particularly if you're an extract brewer – it really isn't as awful as it might seem.

If you've reached a boil already, go ahead and add your extract. In fact, many brewers only add extract in the last 10 minutes of the boil – long enough to sanitise the extract, but not long enough to cause excessive caramelisation from boiling all that sugar. Take care when adding extract into boiling liquid, as the liquid may boil over, or splash onto and burn your skin. In addition, if you add the extract too quickly, it may sink to the bottom and scorch.

The only time you should not add your extract is post-cooling, at least not without some amount of processing. Once wort is cooled, utmost care should be taken to avoid contamination with any microbes, and while extract is often packaged and sealed very well, there is always the small chance that bacteria or mould spores will hitch a ride into your fermenter.

If you must add extract into the fermenter, take the time to dissolve it in water and boil for at least 10 minutes, then cool it before adding. This will help dissolve the extract into water, and make it easier for the yeast to metabolise it. It will also help protect you against infection risk.

DRY VS. LIQUID MALT EXTRACT

Dry and liquid malt extracts are different but equal means of attaining a malt base for your homebrew. Dry malt extract tends to have a longer shelf life than liquid malt extract, but both are excellent base ingredients for making beer.

It is important to always stir the liquid while adding malt extract to the boil to avoid scorching or boil-overs.

24 I'm not sure how to use grain in my extract brew

CAUSE

Partial-mash recipes often come with precious little protocol beyond how much grain to use, leaving the brewer to make assumptions about how to use grain in any given recipe.

SOLUTION

Most extract recipes are actually partial-mash recipes – containing a few pounds of grain for added character – but most don't give much consideration to the content of grain, aside from adding colour and a bit of extra flavour.

It's often best to consider partial-mash recipes as a 'mini-mash'. If grain steeps at the correct temperature – 65°C–67°C (150°F–154°F) – any enzymes in the barley will activate and begin to break starches down into sugar. While the grain is floating in liquid, any residual sugar – or any sugar that is formed from starches – will begin to dissolve.

To best add grain into your extract brew, steep grain for at least an hour in as little liquid as possible – try roughly1.4 litres (1.5 quarts) per pound of grain. Consider adding some base malt into the partial-mash grist, in addition to speciality malts, to contribute enzymes and help break any available starches down into sugar.

When removing grain from the wort, allow time for liquid to run out of the grain – it's all valuable sugar! Consider dissolving your extract into the water that has drained out of the grain bags before heating, so that enzymes can work on breaking down any complex sugars that have been introduced by the extract. If possible, pour warm water – 77°C (170°F) – over the grain to help extract sugar.

No matter what, take your time! Extract brewing is sometimes thought of as a shortcut around mashing, but using the principles of a good mash while making an extract brew can produce superior beer.

MUSLIN BAGS VS. NYLON BAGS

Homebrew suppliers often stock both muslin bags and drawstring-enclosed nylon bags for use in homebrew. Muslin bags can be considered disposable. They are soft and stretchable, and sometimes need to be doubled-up to stop grain from slipping through on particularly low-quality bags, but they're inexpensive and can easily be sliced open to dump grain, or added directly to a compost heap. Nylon bags are more expensive but are often reusable.

In this picture crushed grain is being added to water in a nylon bag. The drawstring at the top means that this bag can be used for multiple brews.

25 How much water do I use in my all-grain mash?

CAUSE

The amount of water used in an all-grain mash can depend on the recipe, and whether there is space in the mash tun for the requisite grain and water. Brewers may not know the best water-to-grain ratios.

SOLUTION

A loose mash – a lot of water – will favour proteolytic enzymes, which break down proteins and denature enzymes, meaning that starch conversion may happen a bit slower in looser mashes. It can also be difficult to form a mash bed in a loose mash because so much of the grain is floating.

A thick mash – not much water – will favour diastatic enzymes, which help break down starches into sugars, meaning that starch conversion may happen quickly, but lautering may be more difficult because proteins have not been efficiently broken down or denatured.

In general, when creating an all-grain mash, the sweet spot is between 1.2 litres (1.25 quarts) of water per pound of grain and 1.65 litres (1.75 quarts) of water per pound of grain. That range allows enzymes to work efficiently from both ends, creating a well-converted mash that allows water to pass through the mash bed, as required for a good run-off.

In certain brewing conditions – big beer in a small mash tun, lack of secondary water vessels, for example – it may be necessary to brew with as low as a 0.75:1 ratio, or as high as a 2:1 ratio. In the end, good beer will still be achieved, though it may require a little more patience. In circumstances where mash ratios are not ideal, consider mashing for a longer amount of time to allow enzymes time to work in the mash bed.

It is important to ensure that there is enough room in a mash tun for the extra water added during the lautering process.

26 I can't get a filter bed to form in the mash

CAUSE

Run-off is too quick, or mash temperature was too low, producing a very cloudy wort or a lot of husks and debris entering the kettle.

SOLUTION

First, be sure that the mash temperature is correct. Unless a speciality recipe is being brewed, aim for between 66°C and 68°C (150°F and 154°F) for most mashes. If you can, heat the mash bed to 77°C (170°F) before run-off. This will arrest enzymatic activity and denature many proteins that may cause a sticky run-off. Otherwise, be sure that the sparge water is around 77°C (170°F). The mash bed will slowly come up to temperature as lauter progresses.

If the grist consists of a large amount of high-protein grain like wheat, rye, or oats consider doing a small step mash. A rest between 37°C (98°F) and 45°C (113°F) will activate the enzyme beta glucanase, which will help break down beta glucan. Beta glucan is a gummy cellulose that can create a cement-like mash. A 20-minute rest should be enough to keep your mash beta glucan-free.

Next step is recirculation. Transfer 5 to 10 per cent of the volume of your mash into a vessel (your kettle or another pot) then return it to the top of the mash. Recirculation establishes a good mash bed by running liquid slowly out of the bed, allowing the barley husks to form a natural filter. It also removes protein and bits of barley husk that may have fallen below the false bottom of your mash tun, and places them on top of the mash bed, filtering them out while retaining the valuable sugar present in first runnings.

Good recirculation, and a slow start to a run-off, can help establish a good mash bed – once wort is free of any visible bits of grain, you can slowly increase the run-off speed.

A good filter bed requires the right temperature mash. This is vital for good sugar conversion, and an easy run-off. It is helpful to keep a thermometer to hand throughout the boiling process to ensure the mash temperature stays consistent.

FALSE BOTTOMS VS. FILTER SCREENS

There are a variety of options for building or purchasing a false bottom for a mash tun, but they all come down to the following choices: a false bottom, or a filter screen. Filter screens can often be easier and less expensive to instal and maintain, but false bottoms are a bit easier when dealing with a stuck mash or an otherwise finicky grain bed.

27 The grain bed collapsed during lauter

CAUSE

Water is no longer running through the mash bed, or proteins present in the mash have caused a 'stuck mash'.

SOLUTION

This is often associated with high-protein mashes with adjuncts that contain a lot of gummy proteins, like wheat, rye or oats. However, a stuck mash or collapsed bed can happen on any beer. First, check Problem 26 for how to establish a good filter bed.

Check the grain bed has enough water in it. In a small mash tun, the top of the liquid in the mash tun should be between 1.3 and 2.5 cm (0.5 inches and 1 inch) above the top of the grain bed. This ensures that the grain floats in the water rather than collapsing against the false bottom. Also check your sparge water is hot enough. You should be sparging at 77°C (170°F).

If you're brewing with high-protein grains, add rice hulls to help create channels through the mash bed. Ideally, these are added before the mash starts, but stirring them into a stuck mash can often help get liquid moving again. Stir them so they distribute throughout the mash bed, the closer to the bottom the better, then recirculate wort and re-establish a filter bed.

Underletting can often restart a stuck mash but is challenging if you don't have robust equipment (or a pump). Push some hot water into the mash tun from the bottom up, floating the mash on a column of sparge-temperature water, then stir the mash bed, scraping any gummy proteins from the false bottom, and recirculate wort to establish a new filter bed.

If this is a consistent problem it may be that your grain is crushed too finely (see Problem 16).

✿ This picture shows recirculation in action. Brewers recirculate wort by extracting 5 to 10 per cent of the volume from the bottom of the mash bed, transferring it to another clean, uncontaminated vessel, and returning it to the mash tun. They do this before starting run-off in order to ensure clear wort and a stable mash bed.

28 My infusion mash won't reach target temperature

CAUSE

Strike water has not been adjusted for the ambient temperature of grain or equipment. Even if 68°C (154°F) water is used, if strike water is only 20°C (68°F), this will not result in a mash temperature of 68°C .

SOLUTION

To correctly achieve mash temperature, it's important to remember that heat will be lost when introducing hot water to cooler grain. While it's easy to use brewing software or an online calculator to find the best target mash temperature, this is also an easy calculation you can do at home. You need to know the number of litres of water per pound of grain being used (L), the current temperature of the grain (C), and the target temperature of the mash (T), and apply it to this equation:

$$\text{Water Temp} = (0.2 \div L) \times (T - C) + T$$

For instance, if you have 1 litre (1 quart) of water per pound of grain, the grain is at room temperature, 20°C (68°F), and your target mash temperature is 68°C (154°F):

$$0.2 \div 1 = 0.2$$
$$154 - 68 = 86$$
$$0.2 \times 86 = 17.2$$
$$17.2 + 154 = 171.2°F$$

However, bear in mind that the temperature of your equipment may also play a role. This equation will allow a ballpark estimate, but if hot water is being added to a cold, metal mash tun, additional heat may be lost. It's always best to keep a thermometer handy to be sure that you've met mash temperature correctly.

🌿 *Temperature management is one of the most important parts of brewing so a quality kettle thermometer is a vital piece of brewing equipmement.*

KNOW YOUR ENZYMES: BETA AMYLASE

Beta amylase is one of the two major enzymes that brewers rely on to break starches down into sugars in the mash. Its optimal temperature range falls between 54°C (130°F) and 66°C (150°F). It attacks the ends of starch chains, breaking them off into a pair of connected sugar molecules called 'maltose', which is easily digestible by yeast. This gives beer a high fermentability, and a thinner, crisper body.

29 My mash won't hold temperature

CAUSE

A poorly insulated mash tun, or a mash tun made of the wrong material, allows heat to escape the mash, leading to poor efficiency or stuck mashes.

SOLUTION

The mash temperature should remain constant through the entirety of the mash to ensure that enzymes remain active to break down all the available starches into sugar. In most cases, the thermal mass of the grain and the water together should retain enough heat to stay relatively consistent through an hour mash, but that is not always the case.

While a mash tun could be any vessel with a false bottom in it, most homebrew mash tuns are made from insulated water coolers which are meant to keep ice and water cold. They are equally effective at keeping a mash warm.

When using a single-walled pot or kettle, which conducts heat much better than insulated plastic, the brewer should monitor temperature to be sure it doesn't dip, and if it does, reheat the mash tun if possible. However, care should be taken not to scorch the mash, but to be sure that heat is transferred as evenly as possible throughout the grain bed.

If a direct heat source is unavailable, a home-brewer can place a copper-coil heat exchanger inside the mash tun, and cycle mash-temperature water throughout the coil to maintain temperature.

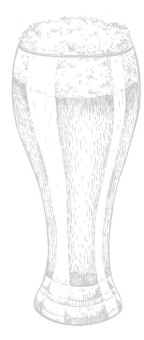

KNOW YOUR ENZYMES: ALPHA AMYLASE

Alpha amylase is one of the two major enzymes that break down starches into sugars in the mash. Its optimal temperature range is between 68°C and 72°C (155°F and 162°F). It attacks random points in starch chains, breaking them down into different-sized sugars, some of which are broken down further into sugars digestible by yeast, and some which aren't — known as dextrins. This gives beer some residual sweetness, and a fuller body.

Brewers need to regularly monitor mash temperature to avoid stuck mash. A stuck mash can mean a very long day, and possibly mean that off-flavours will be extracted from grain.

30 My mash bed keeps running dry

CAUSE

Poor flow management during lauter can empty a mash tun too quickly, collapsing and compacting the mash bed and reducing efficiency.

SOLUTION

Lautering is relatively easy and should be a balance between temperature and time. When lautering, the water level should always be just an inch or so above the level of the grain bed, and the grain bed should never run dry.

Since the purpose of lautering is to extract sugar from the barley in the vessel, it's important that the water has enough contact time with the barley to dissolve that sugar. When all else fails, close the run-off valve on the mash tun, then reopen it slowly so that a mere trickle runs out. On a home-brewing scale, it should take about 30 minutes to run off 15–19 litres (4–5 gallons) of wort, or about 1 hour to run off 38 litres (10 gallons) of wort.

Some brewers lauter using a technique known as 'batch sparging'. Batch sparging is the act of topping off the mash with sparge-temperature water, running off until the mash is dry, then filling the mash tun with sparge-temperature water again, repeating until volume is met. In this case, conventional wisdom is that there is no reason to run off slowly, since all the grain will meet hot water again, but a slow run-off may lead to higher efficiency and cleaner wort.

BATCH SPARGING

Some brewers say that batch sparging is a faster way to lauter, with no impact on efficiency. It can be a beneficial method when space is tight. Simply heat all the lauter water you need to 77°C (170°F) and fill the mash tun with it. Vorlauf until the wort runs clear, then empty the mash tun completely. Refill and repeat until boil volume has been reached.

Allowing the mash bed to run dry during run-off can collapse the grain bed, making it very difficult to then lauter.

31 I don't know when to stop lauter

CAUSE

It can be difficult to know intuitively when to stop lautering. There are several schools of thought to follow, none of them perfect.

SOLUTION

There are three ways to figure out when to stop lautering, each with their own issues:

- Finish lauter when the wort in the kettle reaches the predicted/desired pre-boil gravity. While this means you will have hit a recipe target, depending on the efficiency of your mash you may run much more, or much less, water through it than needed, leaving you with much more, or less, wort than you intended. Remember that wort layers easily, with the densest wort at the bottom of the kettle and the least dense wort on top. Mix the kettle thoroughly before taking a gravity measurement.

- Finish lauter when the gravity of the run-off water reaches 1.008. In theory, the likelihood of extracting tannins from the husks of the barley, rather than just sugars, increases below this point. There's still a risk of lautering more, or less, wort than is needed. Unless you're brewing an extremely low-gravity beer, the likelihood of tannin extraction is low.

- Finish lauter when you have the correct amount of liquid in the kettle – the size of the batch with extra wort to make up what you will lose in evaporation. While you may not always hit your gravity targets perfectly, you will have achieved one of the important goals: to have the correct batch size. If you maintain a consistent level of mash efficiency, this third option is an excellent method, particularly for creating a specific amount of re-creatable beer.

🦋 *The alternative to batch sparging is fly sparging. Fly sparging is a method used by most commercial breweries and all-grain home-brewers. Using a simple device such as the one pictured here can make fly sparging at home an easy possibility.*

EVAPORATION RATE

As a rule, home-brewers will see 6 to 8 per cent of their wort evaporation over the course of an hour boil, depending on liquid volume and heat source. Brewing on a hob is less vigorous so evaporation rate is rarely high. However, brewing on outdoor propane burners can sometimes boil as much as 10 to 12 per cent in an hour. Take measurements of volume before and after boil to calculate your evaporation rate.

32 My mash temperature was way too cold

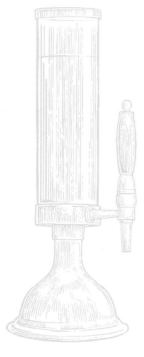

CAUSE

Strike water was not warm enough to achieve target mash temperature. Mash is cooler than anticipated, leading to poor conversion, loss of efficiency or a thin-bodied beer.

SOLUTION

The most efficient starch conversion in modern mashes falls between 66°C and 68°C (150°F and 154°F). If your mash temperature is too low, the easiest solution is to add more hot water to the mash. However, given the size restrictions of your mash tun this may not be possible. It takes much more energy, thus more water, to raise the last few degrees of a mash. It may require the addition of very hot, even boiling, water. To find out how much water to achieve a specific temperature, you need to know the desired temperature (T), the temperature of the water you will add (W), the current temperature of the mash (M), the amount of grain in the mash (G), and the litres of water currently in the mash (L).

Amount of Water to Add = (T − M) × (0.2G + L) ÷ (W − T)

If you have a 60°C (140°F) mash, and you'd like to achieve 68°C (154°F), using water at 93°C (200°F), and there's 5.4 kg (12 lb) of grain and 11.4 litres (12 quarts) of water:

$$(154 - 140) \times (.2 \times 12 + 12) \div (200 - 154)$$

Or

$$14 \times 14.4 \div 46 = 4.4 \text{ quarts of 200-degree water}$$

If there is not enough room in the mash tun, it may be necessary to run wort off, carefully bring it to a boil, and re-add it to the mash in a process known as 'decoction mashing'. The same calculation applies, but bear in mind that the drawn-off liquid is no longer in the mash.

33 My mash temperature was way too hot

CAUSE

Strike water was too hot to achieve target mash temperature. Mash is warmer than anticipated, leading to poor conversion, loss of efficiency or wort with low fermentability.

SOLUTION

If your mash temperature is too hot, you must act quickly to reduce the temperature, or you run the risk of denaturing enzymes that are important to the mashing process. Enzymes do not denature instantly, but the warmer the mash, the faster the denaturing process.

If you are a few degrees off from your target, it may be more efficient to stir the mash vigorously, bringing mash from the middle of the bed to the top. This releases energy from the mash, as steam and heat rises from the top of the mash bed, and can easily reduce the overall mash temperature by a few degrees.

If your mash temperature is much higher than intended, you may need to add cold water, then proceed with the same formula used to increase mash temperature (see Problem 32).

Amount of Water to Add = $(M - T) \times (.2G + L) \div (T - W)$

Thus, if you have a 77°C (170°F) mash, and you'd like to achieve 68°C (154°F) using water at 16°C (60°F), and there is 5.4 kg (12 lb) of grain and 11.4 litres (12 quarts) of water:

$$(170 - 154) \times (0.2 \times 12 + 12) \div (154 - 60)$$

Or

$$16 \times 14.4 \div 96 = 2.3 \text{ litres (2.4 quarts) of 16°C (60°F) water}$$

If there is not enough room in the mash tun, it may be necessary to chill water using ice (or even add ice to the mash bed) to achieve a significant temperature drop without overflowing the mash tun. As always, be sure that temperature is balanced throughout the mash bed.

34 What do I do with my spent grain?

CAUSE

After mashing in, spent grain – the wet barley leftover from the mash – can be a major waste product that is difficult to dispose of.

SOLUTION

Think of spent grain as a by-product rather than a waste product. There are many alternatives to throwing away spent grain. In many areas it is frowned upon to have large amounts of biodegradable waste. Commercial brewers can send their spent grain to a farm to be used as animal feed, but at home you have alternative options.

- If you have the space, spent grain makes excellent compost that breaks down quickly because of its high fibre and high moisture content. Make sure it's aerated and mixed in well so that it doesn't get smelly.
- A little spent grain can add a lot of flavour and texture to homemade bread. However, it has a lot of moisture and not much in the way of gluten, so use sparingly or it may stop your bread from rising.
- Homemade biscuits can be easier to make than bread because they don't rely on rising action.
- Mix in a little peanut butter and sweet potato to make nutritious and delicious treats for your dog.
- Some intrepid brewers use spent grain as a medium for growing mushrooms (using proper pasteurisation and treatment).
- Mixed with paper pulp or sawdust, compacted, and dried, spent grain can be used as a source of heat.
- Give your waste to a local farm for their compost, or as a supplement to their animal feed – it may net you a few vegetables in trade if you're a good negotiator!

Whatever you do, always do proper research and preparation to ensure safety.

THE NUTRITIONAL COMPOSITION OF SPENT GRAIN

Spent grain is just barley that's been cracked open and had most of the sugar, starch, carbohydrates and some of the proteins removed. It is high in dietary fibre, polysaccharides, vegetable protein and even calcium.

🌸 *Spent grain has many uses but it can also be very hot immediately after run-off, so care should be taken when handling it.*

CHAPTER FOUR
BOILING

Boiling is vital to making beer. First, it allows the brewer to be sure that the liquid that they're creating is virtually microbe-free. Very few microbes can survive a boiling. Later on, when brewers add a very specific organism into the liquid, they want to be sure what they are adding is the only one in there.

Boiling provides a means for driving off some volatile off-flavours that can derive either from the ingredients or from certain chemical reactions that happen in wort when heat is added. A good rolling boil allows the brewer to remove some chemicals that would otherwise be unpleasant in the finished product.

Hops must be boiled in order for them to go through the isomerisation process. Boiling is also how hop oils are efficiently extracted into wort, providing both bitterness and hop flavour and aroma. Finally, boiling provides caramelisation, sometimes in light amounts and sometimes in greater ones, offering an enjoyable, rich complexity to your beer, which pairs beautifully with roasted or grilled foods.

Problems in the boil usually derive from poor temperature control or measurements – both of which can be solved through careful attention to the necessary processes.

35 My boil has long stringy debris in it

CAUSE

As wort boils, proteins and carbohydrates coagulate, making long stringy-looking protein chains. This is known as 'hot break' and occurs with varying intensity, depending on wort composition and temperature.

SOLUTION

Hot break in and of itself is not a problem, though it can appear to be one when first experienced. A lot of hot break can often lead to a clearer beer in the end, because as proteins stick together, they settle out of solution and are less likely to be passed along to fermentation.

To avoid an excessive amount of hot break, cut down on protein content in your wort. This will often come in the form of high-protein adjuncts, such as wheat or rye, but can also be present in some barley. Six-row barley has a higher protein content than two-row barley, and even among those, different batches of malted barley may have higher protein content than others. Ask at your local homebrew shop if they have spec sheets, provided by the maltster, with details of protein content.

Hot break isn't something you necessarily need to avoid or fix, it is just something that you want to avoid bringing into your finished wort. In order to help hot break settle out of solution, use a fining agent, such as Irish moss, to help further coagulate the proteins.

At the end of a boil, stir the wort in your pot into a whirlpool and allow the whirlpool to settle before transferring wort into your fermentation vessel, siphoning wort from the edge of the pot, leaving all the settled proteins, carbohydrates, hops and whatever else you put into your boil in the centre pile of trub.

WHAT IS A FINING AGENT?

A fining agent is a chemical that is added to beer (or wine) to aid in the precipitation of haze-forming compounds, which are primarily proteins. The fining agent will typically be made up of large, positively charged molecules that will stick to proteins in the solution, adding weight and allowing those proteins to settle out of solution more quickly.

A good rolling boil has many benefits; for example: caramelisation, evaporation, isomerisation, sterilisation, and driving off unwanted off-flavours.

36 My boil ended prematurely

CAUSE

The power went out, the propane tank emptied, or something else unplanned happened to cause the boil to end prematurely.

SOLUTION

Recovering from a loss of active heat in the middle of a boil can be really tricky, but not impossible.

First, make note of how long the wort was boiling to best make decisions.

- Less than 10 minutes: your wort has pasteurised, but not necessarily sterilised. Hops have not had a chance to isomerise and create bitterness. Off-flavours may remain. Find a way to resume boil, and boil again for 60 minutes.
- 10–30 minutes: your wort has been sterilised and some isomerisation has occurred. The beer will not be as bitter as planned, and may be missing some planned caramelisation qualities. If resuming boil at some point soon is possible, do so. You may consider adding late-addition hops and steeping before cooling.
- 30–45 minutes: wort has been sterilised and most isomerisation has occurred. The beer will not be quite as bitter as planned, but it may not be a detectable change, depending on the style. Pitch late-addition hops, cool the wort, transfer to fermenter and pitch yeast.
- 45–60 minutes: you can consider this boil complete and proceed as normal.

There are instances of 'no-boil' or 'raw' beers made by some home-brewers. These are often beers soured with lactobacillus bacteria, with low alcohol content and little-to-no bitterness. Raw beers have the risk of producing off-flavours and could be subject to much different bacterial infections than beer that has been boiled. Proceed with caution if you cannot boil your beer.

❧ *Brewing outdoors on a propane burner rather than indoors on a hob can mean a significantly faster and less messy brew day. Ensure that your propane gas tank is sufficiently full and in good working order before starting your boil.*

37 I don't have enough wort

CAUSE

At the end of a boil, the final measured volume is below what is expected. Brewer did not use enough water in lauter or steep, or experienced a higher than expected rate of evaporation.

SOLUTION

In the worst-case scenario, there is less beer than planned at the beginning of the boil, which will likely yield a higher gravity than expected at the outset.

In extract brewing, it's common to have less wort than the final fermentation and it is even something that can be planned for as part of a 'partial boil' (See Problem 70). It is typical and expected to add water to the wort to achieve the correct volume.

In all-grain brewing, if the problem was not lautering for long enough you may have a much more concentrated wort than originally planned. It is better, of course, to lauter the correct amount of liquid the first time around, as any correction will no longer produce the beer as exactly specified by the recipe, and bitterness may be affected. The best solution is to take a gravity reading once the wort has cooled and then add a small amount of distilled water. Continue taking gravity readings and adding water until the target gravity is achieved.

If the problem is that your rate of evaporation was higher than expected, you should be able to add water to achieve the target volume.

It is ultimately up to the brewer to decide whether the best course of action is to fix this problem on the fly, or just be sure to correct the next batch.

A small wort, or concentrated wort, is not necessarily a bad thing, depending on what style of beer is being made, and if you have a plan for what to do next.

38 I have way too much wort

CAUSE

Inaccurate measurements or inattentiveness leads to a kettle which has much more wort than anticipated, which may produce a watered-down flavour, boil-overs or an unexpectedly low alcohol content.

SOLUTION

First, measure the gravity of the wort and compare the result to what is expected in the pre-boil gravity of the recipe. There is a chance that a very efficient mash has produced more sugar than anticipated, so gravity targets can still be hit even with a higher volume. If the liquid can fit into the kettle or into the fermenter, there is no reason to worry about additional volume.

However, if you are low on gravity, or you have limited fermenter space, the easiest solution is to boil longer to reduce the volume of your wort. Most home-brewers will see an hourly evaporation rate of 6 to 8 per cent. If you have 26 litres in the kettle, and your goal is to achieve five, you may need to boil up to two-and-a-half hours or longer. Be sure to have a means of measuring liquid within the kettle (see Problem 78) and measure frequently throughout the boil, bearing in mind that boiling liquid takes up more space than non-boiling liquid.

Remember that hop additions are generally documented in the 'time before the end of the boil' notation of a recipe. If you are boiling for an additional time, you may want to wait before your initial hop addition to avoid excessive bitterness.

Bear in mind that longer boil times will lead to more caramelised wort, meaning a darker beer with a higher portion of unfermentable sugars and in some cases a more caramel character. Some styles, like Scottish ales, are brewed in this manner on purpose, but most are not.

MAILLARD REACTIONS

A Maillard reaction is a chemical reaction between an amino acid and a reducing sugar in the presence of heat. It is the reaction that causes things to 'brown' while cooking, causing caramelisation, and often creates tasty flavours at the same time. As beer boils it is constantly undergoing Maillard reactions, as sugar in the wort caramelises. Similar to the cooking process, the presence of these 'browning' characteristics is one of the reasons beer pairs so well with roasted and grilled foods.

Avoid boil-overs by ensuring your brew pot is large enough to hold your wort plus an extra 25 per cent of space. Wort takes up more space when it's boiling. Depending on the gravity reading an excess of wort can be reduced by boiling for a while longer, but this might cause unwanted flavours in the final beer.

39 After boil, my beer won't cool down fast enough

CAUSE

Inefficient heat exchange may mean long cooling times, reducing cold break, leading to a cloudier beer, and increasing the risk of DMS formation or contamination.

SOLUTION

The first method that most home-brewers learn to cool their wort is to rest their kettle in a cold-water bath, which helps draw the heat of the wort into the water, thereby cooling the wort. This can be an effective means of cooling homebrew, but can often feel a bit labour-intensive. When a kettle is submerged in a water bath, heat is transferred directly to the water around the kettle. However, once that water gets warm, it can no longer transfer as much heat, so the water on the outside of the kettle must be kept in motion to continually bring new, cool water in contact with the kettle. At the same time, the wort on the outside of the kettle cools first, meaning the hot wort in the centre must be moved out toward the sides of the kettle. The water-bath method is usually fastest when both the wort and the water bath are being stirred.

In some warmer climates, heat exchangers such as copper-coil exchangers or plate chillers may be inefficient due to the ambient groundwater temperatures. Although an additional expense, a second heat exchanger may be the best solution. It can be used additionally to cool wort or even to cool the water on the way to the heat exchanger by using an ice bath or other cooling method on the secondary heat exchanger.

MAKING COPPER COILS CHILL FASTER

Plopping a copper-coil wort chiller into a kettle and running cold water through the coils will eventually cool wort down to temperature, but, like a hot kettle in a bath tub filled with water, it often helps to stir. If the connections on your copper coil are robust enough to handle the movement without dripping, move the copper coil slowly from side to side in your kettle while cooling is happening, to keep hot wort moving across the cold coils.

An ice bath is a simple and cheap method for cooling wort. If using an ice bath keeping the liquid inside moving will ensure that the cooling process is quicker and more efficient.

40 I keep having boil-overs

CAUSE

During boil, proteins in the wort form a dense foam that is lifted by gas escaping the liquid, causing liquid to boil over the side of the kettle.

SOLUTION

The most cost-effective way of avoiding boil-overs is temperature control, and manually breaking surface tension with a spoon or other stirring device. Stirring the foam back into the liquid can often help stop it from boiling over the side. Alternatively, quickly lowering the temperature (or removing the kettle from direct heat) can reduce the amount of gas escaping and stop the wort from boiling over.

Always be careful when adding hops. Adding hops before the wort comes to a boil can sometimes add enough hop oil to help break the surface tension that causes boil-overs. However, adding hops after the boil has begun can often create a layer of hops that acts much the same as the layer of foam at the beginning of a boil, and can cause boil-overs.

Commercial anti-foam products are available to home-brewers – ask about these at your local homebrew shop. They usually come in the form of a silicon-based emulsion that works by breaking surface tension in the boil. If the usage rate is too high, these products can lead to residual bitterness in beer, but used correctly they have no effect on the end product.

Finally, if boil-overs are a frequent problem, you may consider purchasing a larger kettle to brew in, or brewing concentrated wort (see Problem 70). Concentrated wort can still boil high, even more than normal-strength wort, but there will be more room in your kettle.

Boil-overs are not just a waste of wort, and therefore a waste of beer. They can also be very dangerous, as boiling, sugar-filled water splashes and erupts over the surrounding surfaces.

BREAKING SURFACE TENSION ON THE BOIL

Blowing across the surface of the boil to dissipate bubbles, or setting up a fan to do it for you, or just stirring frequently, can break surface tension for long enough for the boil to stabilise. Adding a small drop of olive oil can break surface tension in the boil and can act as a source of lipids for yeast, but be careful as too much may cause head retention problems.

41 I forgot to add my hops

CAUSE

Forgetting to add hops on time may lead to a general lack of hop flavour (overly sweet or lack of bitterness). Hops are an active anti-bacterial so, if not added at all, there is a slightly increased risk of infection and shorter shelf-life.

SOLUTION

If the boil has not yet ended, there is still time to add hops. Hops added close to the time listed in the recipe should be 'close enough' for most cases. There will be a slight difference in isomerisation and hop utilisation, but not enough to be detected by the majority of drinkers in a side-by-side comparison with another batch of the same beer.

If the addition time has passed by more than 10 minutes, then you must decide whether you are okay with less hop bitterness/less hop flavour. If not, consider resetting the timer on the boil from that point. If you've missed your 45-minute addition and it is not time for your 15-minute addition, make the addition and time 45 minutes for the rest of the boil. This will cause additional bitterness contributed by the previous hop additions.

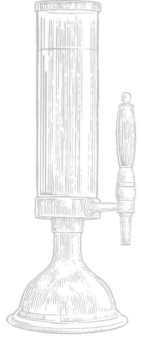

If the boil has ended, there are two options. Hops are vital for beer to taste like beer, and they need to be boiled to create bitterness. You may boil hops in a small amount of water for the desired time (e.g. 60 minutes, 45 minutes) and add the resulting liquid to the fermenter after it has been cooled.

The last option is simply to add the hops as a dry-hop addition. In this case hop flavour and hop aroma will be added (though different in character from boiled hops), but bitterness will not.

GRUIT: BEER MADE WITHOUT HOPS

Before hops were widely used, humans made their beer bitter using a wide array of herbs, including mugwort, wormwood, and other very bitter – sometimes poisonous – herbs. Modern home-brewers sometimes make gruits, which are malty beers made with alternative, traditional bittering agents and a wide variety of spices for flavouring. They can be fascinating brews to try, but make sure all the herbs used are safe for human consumption!

Hops are an important addition to a brew due to their anti-bacterial properties. If a brewer forgets to add them at the time stated on the recipe there are options available to rectify this, but the final beer will likely end up different from what was originally intended.

42 I added more hops than I was supposed to

CAUSE

Inaccurate measures, or a simple mistake, leads to more hops being added at a particular time than was called for in a recipe.

SOLUTION

If the boil has already ended, there is nothing you can do to remedy this issue. You now have a beer that will have a more bitter, hoppy flavour or aroma.

If the boil is still going, there is a decision to be made: go ahead with the higher bitterness derived from the hop addition (i.e. do nothing and see how it turns out) or adjust the rest of the beer going forward. There are two possible adjustments to be made, and they can be made together or separately.

First, you can reduce the boil time. Since more alpha acids isomerise the longer the boil lasts, if you reduce the boil time you can reduce the bitterness derived from the hop addition. However, there will be a certain amount of extra hop character in the beer regardless, and you may also run the risk of not completely driving off potential off-flavours during the boil.

Second, you can choose to reduce later hop additions. Since all hop additions add a certain amount of bitterness to the wort, reducing late hop additions can help you keep bitterness in check, though you will do so at the risk of flavour and aroma characteristics.

You may choose to reduce the boil time but add later additions at their full amounts, thereby reducing bitterness from boiling but retaining later hop characteristics.

Ultimately, this is one of the many cases where brewing is an art as much as a science and the decision about how to deal with a circumstance is in the brewer's creative hands.

🌺 *Adding hops is as simple as dropping pellets into the brew.*
Do so carefully, however, as adding hops too quickly may
sometimes lead to boil-overs.

CHAPTER FIVE
FERMENTATION

At the base level of beer-making, a brewer adds yeast to a sugary liquid, and the yeast eats the sugar to create alcohol, ethanol and flavour. It's often said that brewers just make wort, and the yeast makes the beer. In many ways, brewers are really just yeast ranchers. The job of the brewer is to create a nutrient-rich environment in which yeast can grow, eat, live and make more yeast. In return, yeast creates ethanol and CO_2 which makes an enormous range of flavours that humans find pleasant. It's a wonderful, happy relationship.

In some ways, yeast cells are much like grazing animals. Give a few yeast cells an enormous pasture and it will take time for them to consume all the food, or they may never get to it all. There is also a risk from other 'grazing' organisms. On the other hand, adding a large amount of yeast cells to a small pasture may mean the yeast cells eat all of the food too quickly – before they have time to settle in and make more yeast.

Most fermentation problems arise from not treating yeast cells well, or not fully understanding their needs and habits. Giving them a balanced environment, the right temperature and a good, steady (but not overwhelming) food source keeps them happy, and happy yeast makes great beer.

43 Fermentation hasn't started

CAUSE

Wort was too hot when yeast was pitched, wort is too cold and yeast is dormant, or yeast was dead when pitched.

SOLUTION

Yeast is a live organism and should be handled with care. While it can survive a wide array of adverse conditions, the job of the brewer is to create an environment where yeast can thrive.

Wort that has not been properly cooled can be fatal for yeast. If yeast is very cold to begin with, it can even be shocked entering a lukewarm environment. On a brew day yeast should be slowly warmed up, as close to pitching temperature as possible. Wort should be as close as possible to 16°C–21°C (60°F–70°F).

In some climates and conditions, it can be difficult to cool wort to 16°C (60°F) (see Problem 39). Yeast will survive when pitched into wort that is as hot as 32°C (90°F), but in these conditions the yeast may create fusel alcohols or grow unhealthily. It is best to make every effort to cool wort to a good pitching temperature.

If the wort is too cold, yeast may go dormant upon being added. In this case, warming the wort (with a heating pad or even just a warm room), rousing the yeast, should start fermentation.

In some unfortunate situations, through poor shipping conditions, shelf-life management or just bad luck, the yeast may be dead. In this case, the best option is simply to pitch more yeast as soon as possible.

FERMENTATION TEMPERATURE

Each yeast has its own ideal fermentation temperature. The manufacturer will specify a range of temperatures that the yeast ferments best in. The colder a yeast ferments, the 'cleaner' it will be. The warmer a yeast ferments, the more flavour compounds it will create, meaning you can control the flavour profile of individual fermentations through temperature control. Ale yeast will ferment best at around 20°C (68°F) and lager yeast will ferment best at around 11°C (52°F).

Accurate temperature readings are vital in managing fermentation. Remember that the temperature on the outside of the demijohn is often several degrees cooler than the liquid in the centre.

44 Fermentation ended earlier than expected

CAUSE

The yeast that was pitched was insufficient (in quality or quantity), or wort conditions changed, or the yeast went dormant, resulting in wort that has a final gravity higher than expected for the given strain and beer.

SOLUTION

If fermentation was started with a commercial pitch of yeast (a vial or pouch), it's possible that there was not enough viable yeast for a robust fermentation, either because the gravity of the beer was too high or because there was something wrong with the yeast pitch. Underpitching yeast can mean that the yeast in fermentation will run out of resources to grow and make more yeast before they've run through available sugar. Put another way, your fermentation ended early.

To be sure that you have live and viable yeast to ferment the wort you've created, it's always best to start with a yeast starter (see Problem 21). If you have a fermentation that has stopped because of a low pitch rate, the easiest and simplest thing to do is to add more yeast. It can be of the same kind or a different, more attenuative yeast; just be sure that you have an active yeast culture.

If the wort has become very warm for some reason, it is likely that the yeast has died. You may restart fermentation by adding more yeast, but it is likely that off-flavours have been created by fermentation at a warm temperature or from the dead yeast.

If the wort has become too cold, the yeast may have gone dormant. Placing the fermentation in a warmer environment may spontaneously begin fermentation again. In some cases, it may be necessary to rouse the yeast by either carefully shaking or agitating your fermentation vessel.

45 Fermentation ended quickly

CAUSE

Though typical fermentations take 4–10 days to complete, under certain conditions the yeast may digest all the sugar available to it in a much shorter amount of time.

SOLUTION

In and of itself, quick fermentation is not a problem and, in fact, it may be a boon. However, rapid full attenuations can be indicative of other problems and should be examined so that those problems can be fully addressed.

Verify that the yeast strain has reached its final attenuation by taking a gravity reading and checking the attenuation percentage range specified by the manufacturer. This information is often listed on the yeast package, but can also be obtained on a manufacturer's website. If the gravity did not get as low as you expected, a different problem may be present (see Problem 42).

Rapid fermentation is often due to poor temperature control. Most yeast ferments best at a cool room temperature of 20°C (68°F). However, yeast creates its own heat in fermentation and even a small 20-litre (5-gallon) batch of homebrew can rise to an additional 4°C–6°C (7°F–10°F). In an uncontrolled environment, it's not unusual for a fermentation to reach 27°C–32°C (80°F–90°F). Particularly warm fermentations can result in high ester production (extra fruity flavour), high fusel alcohol production (nail polish, paint thinner or others – see Problem 88), or even yeast autolysis (soy sauce flavour – see Problem 90).

An over-fast fermentation may be indicative of an alcohol-tolerant bacterial infection in fermentation. In this case, the resulting beer may be sour or smell unpleasant. Fortunately, none of these conditions are harmful to the drinker, but it is unlikely the beer will taste as it was designed to.

46 Fermentation seems to last for ever

CAUSE

Too little or unhealthy yeast was pitched, fermentation temperature is too cold, or a wild yeast infection leads to a fermentation that seems to go on for an extremely long time.

SOLUTION

Take a gravity reading on the beer whether the beer is taking a long time to ferment in general or is fermenting much longer than expected. If the gravity is still high, suggesting that fermentation is slow, check the temperature of the space where fermentation is taking place.

Yeast can create its own heat during fermentation, but if the ambient temperature of the fermentation space is too cold, the wort may cool enough for yeast to go dormant or arrest fermentation. In this case, placing the fermenter in a warm area may easily restart fermentation. You may need to rouse the yeast for this to happen.

If the gravity was still high, and the fermentation environment is a reasonable temperature, then it is likely that the yeast that was pitched was unhealthy or there was not enough of it. In this case, it is best to pitch more of the same yeast to help restart fermentation as soon as possible to avoid contamination.

Finally, if the gravity was much lower than expected but fermentation still appears to be occurring slowly, it's possible that there is a wild yeast infection. Wild yeasts are often much more efficient fermenters than domesticated saccharomyces strains and may slowly work through all available sugar. The best way to know is to taste: If there are undesirable characteristics – usually harsh phenolics (smoky, plastic or medicinal flavours) arising from wild yeast contamination – then you may need to consider dumping the batch.

HEAT-LOVING YEASTS

There are a few Belgian farmhouse yeasts that famously ferment well in the 27°C–32°C (80°F–90°F) range, which is far higher than other yeasts. Some home-brewers heat their fermentation to those temperatures, but the fermentation — an exothermic reaction — will create its own heat. If you apply heat to these already very warm fermenting yeasts, you risk off-flavour formation and possibly killing the yeast.

A ring of spent krausen has remained at the end of this fermentation. If the fermentation on a brew seems to be taking for ever there could be a number of reasons for this.

47 My fermentation smells funny

CAUSE

Fermentations can off-gas undesirable aromas akin to rotten eggs or burnt matches. Some people find the normal yeast characteristics off-putting when concentrated in gas during fermentation.

SOLUTION

For the most part, fermentation smells a little funny no matter what. Many of the flavours that yeast produce in beer are off-gassed during fermentation due to the extraordinary amount of CO_2 that yeast produces during fermentation.

Some Belgian yeasts, Hefeweissen yeasts and lager yeasts have a strong sulphur smell when fermenting that can be alarming, but is totally normal. On the other hand, there are some aromas that can be warning signs for other problems. Watch out for fermentations that smell like rubbish, sewers or goaty earwax. These can be indicative of the presence of organic acids that are a result of some unpleasant bacterial infections.

Some home-brewers use blow-off buckets rather than airlocks. Unless blow-off buckets are cleaned on a regular basis, they can be the cause of unpleasant aromas, as well. It's a good idea to keep a blow-off bucket filled with fresh sanitiser. If any yeast or wort blows off the fermentation and into the bucket, replace the water in the bucket. If the water in the bucket is a good place for bacteria to grow, and there is a trail of krausen up the blow-off hose to the beer, there is a small chance that bacteria may use that pathway to contaminate your beer.

FERMENTATION BY-PRODUCTS

Yeast is an incredible organism that consumes sugar, turning it into ethanol and carbon dioxide, but it's also responsible for up to 900 different flavour compounds in beer. Each strain, with different wort compositions and fermentation temperatures, can create its own unique palette of flavours and textures in a beer. Learn them through experimentation: split wort into separate fermentation vessels, then pitch different yeasts into each one to observe the differences.

These two demijohns each show healthy krausen in the middle of fermentation. The aroma emitted from the airlocks can tell the story of what's happening inside.

48 My fermentation has a slimy white film on top

CAUSE

Particular strains of wild yeast or Brettanomyces may form a white film at the top of a fermentation called a pellicle. Some strains of bacteria may form a ropy slime in fermentation.

SOLUTION

You can either embrace this problem and hope that the flavour characteristics of the wild yeast or bacteria are pleasant, or throw the beer away and start over. If primary fermentation has completed using regular brewing yeast, the wild yeast or bacteria might not be harmful, but the beer may not taste good.

To avoid this issue, it is important to check your equipment for possible contamination. The most likely places for wild yeast or bacterial contamination are inside your fermenter, transfer hose, rubber stopper, airlock, or any piece of equipment that has touched the wort. Before using any of this equipment for another batch of beer, be sure that everything has been fully submersed in cleaning solutions, scrubbed and thoroughly sanitised. If possible, replace soft plastic or rubber parts.

The other potential source of contamination is post-fermentation additions. These often come in the form of additional flavourings, ingredients, or even hops. Since brewers are often using agricultural products to flavour homebrew, there is a high possibility that these items are contaminated with wild yeast or bacteria, particularly if they are freshly picked or foraged. When using such ingredients in finished beer, it is best to sanitise the ingredients before adding them to a beer, either by pasteurising them or by soaking them in a no-rinse sanitising solution.

Freezing ingredients will not kill wild yeast or bacteria. They will merely go dormant until you add them into a nutrient-rich, warm environment, like beer.

The pellicle here can be good news in a purposefully wild fermentation but could spell disaster to a clean brew.

WHAT IS BRETTANOMYCES?

Brettanomyces, literally 'British fungus' (and often just 'Brett'), is the genus of a set of species of sugar-eating fungi. They are similar to regular brewing yeast in that they metabolise sugar and excrete CO_2 and alcohol. Originally identified as a source of infection in British ales, they are still considered an infection yeast in many cases. However, when used intentionally they can create a wide range of pleasant complexes.

49 I'm not sure how to do lager fermentation

CAUSE

Lagers differ widely from ales during fermentation, due to their temperature restrictions, and have special needs that might not be intuitive to brewers.

SOLUTION

Lager yeast, *Saccharomyces Pastorianus*, can be finicky and difficult to deal with. The fermentation temperature range of lagers is normally around 10°C–13°C (50°F–55°F). After bringing to a boil, reduce the temperature of the wort as much as possible. Wort temperatures below 16°C (60°F) are ideal, and lagers need to be fermented in a temperature-controlled chamber to avoid rising above 13°C (55°F).

Lager fermentation can take a long time, and will not produce many of the same visual cues as ale fermentation. Lager yeast does not bind together as well as ale yeast does, so there will not be a large krausen at the top of the liquid where yeast colonies have risen to the top on CO_2 bubbles.

As lager fermentation ends, it needs to be given a diacetyl rest. The purpose of this rest is to allow diacetyl to be formed in the liquid and for the yeast to subsequently digest the diacetyl (see Problem 82). A diacetyl rest is usually achieved by allowing the beer to slowly rise to room temperature (aim for 20°C/ 68°F) over the course of two to three days at the end of fermentation.

Finally, lagers need to be lagered or 'stored' cold. Slowly lower the temperature of the beer to as cold as possible without freezing, then hold the beer at that temperature. Lagering should last 14 days at minimum, and can continue for weeks or months with no detriment to the finished beer, if cleaning and sanitation procedures have been followed.

STICK-ON THERMOMETERS

Most homebrew supply stores will offer stick-on thermometers for the side of your fermenter that can give you an important source of information. Keep in mind that the temperature inside the fermenter will often be a couple of degrees warmer than the room that it's in, and the centre of the fermenter sometimes a degree or two warmer than that.

Most home lagers are made using fermentation chambers built from old refrigerators. With the shelves removed there is plenty of space inside.

50 The liquid from the airlock got into my beer

CAUSE

When lifting a plastic bucket or demijohn, the sides flex causing a slight vacuum, or as wort cools air inside the fermenter contracts, sucking in the contents of the airlock, creating a contamination risk.

SOLUTION

It can be scary to finish a brew day, put wort in a fermenter, and pick the fermenter up by a handle just to watch the contents of the airlock get sucked into the wort that you painstakingly created. In this case, an ounce of prevention is worth a pound of cure.

Be sure that your airlocks are cleaned and sanitised before each use just like the rest of your brewing equipment. Before placing the airlock on the fermenter, make sure it's filled with a sterile liquid and, preferably, one that won't cause damage if it does happen to get sucked back into the fermenter; distilled water or food-grade sanitisers are both good options. Some home-brewers have been known to fill airlocks with cheap vodka, knowing that if it does get sucked back into the fermenter it may add a bit of ethanol but is unlikely to harbour bacteria.

If you haven't taken these precautions and liquid has been sucked back into your fermenter, it is time to treat your homebrew with special care. Be sure you pitch yeast as quickly as possible and watch fermentation for signs of an infection. In many cases, once yeast is in robust fermentation it will outcompete most bacteria for resources and create ethanol which many bacteria cannot live in.

Be ready for signs of the worst, but hope for the best.

CARRYING FERMENTERS

Take extra care when carrying fermenters, particularly glass demijohns. A 20-litre (5-gallon) fermenter weighs 18 kg (40 lb), and that's just the beer. Glass demijohns are often heavy all on their own, slippery when wet, and also breakable. It may be worth ordering a nylon strap harness for your glass demijohns. A broken or shattered demijohn full of beer is not a mess anyone wants to clean up.

S-curve airlocks are inexpensive and easy to use, but be sure that the liquid inside is sterile or sanitised, just in case something goes wrong.

51 Fermentation overflowed my fermenter

CAUSE

Fermentation was particularly active or fermenter was exceptionally full. Krausen overflowed the fermenter and is leaking out.

SOLUTION

It's always a mixed reaction when a fermentation overflows. On one hand, it's nice to see strong, robust fermentation. On the other hand, there is a big mess to clean up and it often seems like there is a lot of liquid loss. An overflowed fermentation shouldn't harm your beer, though cleaning up the mess immediately is the best way to mitigate any potential risk.

The best way to avoid an overflowed fermentation is to leave enough headspace for a robust krausen. In most cases, 25 per cent headspace should be enough to do the trick. Still, some yeasts stick together a little better than others and on some high-protein or high-gravity beers there can still be an issue. If you think you may be in danger of an overflowed fermentation, consider using a blow-off tube rather than an airlock, as it is much easier to clean the inside of a blow-off bucket than the outside of a full fermenter.

Finally, your local homebrew shop will likely carry anti-foaming agents. These are normally used to stop boil-overs in brewing, but can be used in fermentation as well. Take extra care with them, however. Some do not dissolve into liquid, but rather float at the top of a fermentation. If care is not used when moving beer, the anti-foaming agents may end up concentrated in package. While it is not harmful, it can taste very unpleasant in high doses.

BLOW-OFF BUCKETS

It's often easy to find a wide-enough diameter hose to fit snugly in the mouth of a glass demijohn. The other end can sit in a 20-litre (5-gallon) bucket filled with water to create a giant airlock that can handle robust fermentations and make cleaning up considerably easier than a plastic s-shaped airlock.

Sometimes an airlock just isn't enough. The end of a blow-off hose needs to be submerged in liquid so that no gases, bacteria or mould spores can get back up the hose.

52 I'm not sure when/how to dry hop

CAUSE

Dry-hopping is often mentioned in brewing literature and recipes, but it can provoke some questions to home-brewers: when and how do you dry hop?

SOLUTION

If you ask five different brewers how to dry hop, you will have five different answers. There are dozens of methods, none of them totally right or wrong, but across all of them the basics remain the same.

Dry-hopping is the act of putting hops into fermenting or fermented beer to add hop aroma and flavour without increasing bitterness in the beer. The easiest possible way to approach it is just to dump hops into the fermenter after fermentation is complete and let it stand for a couple of days. Rack the beer off the hops and yeast when finished.

A few other methods to consider are:

- Adding hops before fermentation is complete. This allows the yeast to interact with the oils in the hops. Some yeasts can metabolise some hop oils and transform them into pleasant aromatic and flavour compounds.
- Adding hops into a secondary fermenter and then racking them into that fermenter after fermentation is complete.
- Some brewers add hops in stages, adding some hops during fermentation, some after, and even more after that, sometimes chilling the wort before final additions.

In all cases, take care when adding anything into your fermenter. Be sure that everything you're using has been cleaned and sanitised, as this is an additional infection point. Also bear in mind that all of those hops will absorb some liquid. Each hop addition means a little less finished beer.

Ultimately, the best thing to do is experiment to find the dry-hopping regimen that best suits your tastes.

Whole-leaf hops are seen here floating on top of the liquid in a demijohn during secondary fermentation.

53 I didn't transfer beer out of primary fermentation

CAUSE

Busy brewers don't always get to their fermentations in good time. Finished beer may sit on yeast for long periods. Under poor conditions, yeast may 'lyse' (burst) releasing unpleasant flavours.

SOLUTION

First, it should be noted that secondary fermentation is not strictly necessary. It can help clarify beer and in cool conditions can help flavour maturation and shelf stability, but it is a step that can be skipped in many homebrew operations.

Some homebrew literature recommends moving finished beer from yeast as soon as possible to avoid the flavour of yeast autolysis. Autolysis can be best described as yeast death. Under poor conditions, the cell walls of dormant or old yeast can rupture, releasing the contents of the cell into solution and creating a meaty or soy sauce-like characteristic. In homebrew this is rarely a problem.

If you keep your homebrew in a relatively cool, temperature-regulated environment you should be able to leave beer in primary fermentation for weeks to months without worry. Bear in mind that the lighter, more delicate the beer, the more likely it is to develop unwanted off-flavours from sitting on the yeast cake. Keep an eye on your brew. If the yeast just looks like it's settled out of fermentation, it's fine. If the yeast begins to change colour over time, you may be running into a problem.

If you know you'll be unable to package or process your beer, but you know that it is finished with fermentation, do your best to keep it in as cool a location as possible (but not frozen). If possible, refrigerate the fermenter. This will help the yeast go dormant and discourage lysing.

WHEN SECONDARY FERMENTATION IS WORTH IT

While secondary fermentation is not always a necessary practice, there are times when it's worth doing, such as the following: when you're adding flavourings to a beer, particularly flavourings that will cause additional fermentation; when beer hasn't cleared up as much as you'd like it to, but fermentation is complete; when you need to store the beer for a long period prior to packaging.

Some hobbyists have many brews active at multiple points of fermentation so that they always have something waiting to be finished.

54 I added fruit and it looks/smells really foul

CAUSE

Active yeast in the beer has eaten the sugar and extracted colour from the fruit naturally, or fruit contained wild yeast or bacteria that has infected the beer.

SOLUTION

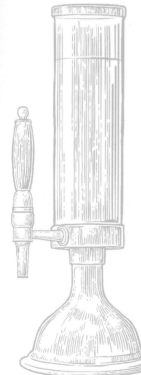

When fruit has been added to fermentation, the brewer should not expect the fruit to continue to look pristine throughout the secondary fermentation process. Yeast breaks down cell walls in the fruit, digests fructose, glucose and other compounds in the flesh of the fruit, and colour from the fruit will dissolve into the beer. Fruit may take on a greyish tint, look mushy, and possibly have yeast mixed in with it. By and large, beer with fruit in it smells overwhelmingly like the fruit that was added. Sometimes fruit fermentations have wine-like characteristics.

There should never be anything that looks or smells like mould in the fermentation. If a white film appears on the top of the liquid and/or the fruit, the beer has been infected by a foreign organism that caught a lift on the fruit into the fermented liquid.

Fruit should be sanitised before being added to finished beer, and it's important to know that freezing does not sanitise fruit. Many wild yeasts or bacteria will merely go dormant in a freezer and will live on to infect beer. If possible, pasteurise the fruit by heating it up to 60°C (140°F) for 10 minutes, or even by boiling the fruit before putting it into the beer; but understand that heat will change the flavour characteristics of the fruit. If you'd like to add raw fruit, wash it well under water, rinse, and soak in a sanitising solution before adding it into the fermenter.

FRUITS THAT WORK WELL IN BEER

There are many tried-and-true fruits for beer. Excellent beers have been made from blueberries, raspberries, blackberries, oranges, grapefruits, passion fruit, pineapple, persimmons, gooseberries, apricots, plums, peaches and dragon fruit, to name just a few. Some fruit loses its flavour and aroma quickly, for example strawberries or kiwis. The best way to find out is to experiment. What fruit/style combinations lie in your imagination?

Fruits may break down an enormous amount during fermentation as sugars and other components are broken down by yeast.

55 I need to measure alcohol after adding fruit

CAUSE

Fruit contains sugar. When added to a fermenter, fermentation will often start up again and while subsequent gravity readings will be accurate, it's difficult to know what the effect of the additional sugar was on the original gravity of the beer, particularly when fruit is added late in the process.

SOLUTION

Some commercially available fruit purees will provide a measure of the sugar contained in the puree, particularly those available through homebrew suppliers. This measurement is often listed in Brix, which is an alternative measurement of density to specific gravity or plato, and used mostly in winemaking. There are many calculators and charts available to translate Brix to your gravity of choice.

When working with fresh fruit, it can be difficult to know how much sugar is being contributed to the solution. One method is to measure the gravity just before adding fruit, then measuring the gravity immediately after fruit has been mixed in to find a new starting point for that period of fermentation.

Bear in mind, however, that some of the sugar may still be locked away inside the fruit itself. Consider that fruits also contain a large amount of water. It is likely that the addition of water will offset the addition of sugar and that you will not achieve a higher alcohol content than originally planned.

Many professional brewers use a refractometer, as seen here, to measure the density of wort. Refractometers are less commonly used by home-brewers, but if used properly they can be a great tool at home as well as in a brewery.

CHAPTER SIX
PACKAGING

Packaging is a necessary and complicated process that allows brewers to enjoy the product that they worked so diligently to create. It can also be the bane of a brewer's life.

Bottles can be finicky and difficult to clean and sanitise. Capping can be a tricky process for a range of different bottles. There is a risk from broken glass, and the process of recarbonating in the bottle to create CO_2 can be dangerous if not done correctly. Bottling can also generally be a messy process.

Kegging is often easier than bottling. However equipment such as kegs and CO_2 cylinders are expensive, and the high cost of necessary serving equipment – a kegerator or keezer – can be a barrier to this step.

Post-fermentation, there will be sufficient amounts of viable yeast to naturally carbonate the beer with the addition of a priming agent, such as sugar or dried malt extract – be that in a keg, pressure barrel or bottle. Most home-brewers start with this method because it's cost-effective and needs no additional equipment. Ale that is naturally conditioned in this way is considered to be real ale.

Packaging problems are often due to cleanliness or pressure problems (or both). At this stage in the process a small mistake can mean the difference between spending an afternoon enjoying a batch of beer or frustratingly opening and dumping bottles.

56 My beer gradually becomes sour in bottles

CAUSE

The beer has a lactobacillus infection. Lactobacillus is a genus of gram-positive, rod-shaped bacteria that can eat the sugar that is left after fermentation, creating lactic acid, which tastes sour.

SOLUTION

A lactobacillus infection in the bottle is discouraging and, once infected, there's no way to recover beer that's been soured. However, you can examine your process to determine where the infection may have occurred in order to mitigate it in the future.

First, is the problem present in some bottles or all bottles? If it is only in some bottles, then chances are the bottles in question were not sufficiently cleaned and sanitised.

If the problem is in all the bottles, then tracing back the route of the beer can help find the issue. Check the bottling apparatus and any hoses. Are there any dead spots where there might be leftover beer, or sticky spots where bacteria could grow? Were they properly cleaned and sanitised before use? During the bottling session, did anything come in contact with the bottling tip that wasn't beer or the bottle that was being filled? Check the bottling bucket or vessel. Are there scratches or cracks inside that might harbour bacteria? Was the vessel properly cleaned and sanitised before beer was introduced? Did anything else come in contact with the beer before it went into the bottles? Did you take a sample out to measure gravity? Was the vessel you sampled with sanitised before touching the beer?

Work backwards from the point where the infection is discovered and investigate each piece of equipment and action. In a worst-case scenario, you may need to replace any soft plastic or rubber parts if they cannot be properly cleaned and sanitised.

Brewers sometimes intentionally use lactobacillus to acidify beer for certain sour styles, but if your beers are gradually souring in the bottle unintentionally it's a sign of an infection."

57 My bottled beer is really flat

CAUSE

If beer has been bottled using priming sugar to ferment and create carbonation in the bottle, there has been a fermentation issue. Alternatively, seals on the bottle caps are faulty.

SOLUTION

Was the correct amount of priming sugar used? Different types of sugar require different dosing rates. An easy rule of thumb for dextrose is half a cup per five gallons. Charts and calculators are available to help determine the amount of sugar you need, and sugar tablets are available for easy measuring. It is possible to dose each bottle with sugar, but be careful not to put too much in, or you may over-carbonate the beer instead.

Are the bottles at room temperature? Bottle conditioning requires re-fermentation in the bottle. Bottles must be kept at a good fermentation temperature. If bottles are kept cool or cold, yeast may be dormant. Keep the bottles at room temperature for a period of time and they should carbonate.

Has enough time passed? Because there are so few yeast cells present in finished, clear beer, bottle conditioning can take weeks, particularly in a high-alcohol beer. Give the beer a couple more weeks to see if carbonation forms, before giving up.

If none of those are the issue, or the bottles were filled from an already fermented beer (out of a keg), check the seal of the crown caps. Caps should fit snugly around the top of each bottle without being over-tight.

If you can pop a cap off with your thumb, rather than an opener, or if you tip a bottle on its side and beer leaks out, you don't have a good seal. Consider re-capping the beers and waiting a few more weeks.

FLAWS IN BOTTLES

Keep an eye on bottles for flaws in the glassware. There will often be bubbles embedded in the glass, and those are okay, but anything that breaks the surface of the bottle that could result in a broken blister bubble is a potential hazard – both as a potential source of glass in beer, but also as an explosion hazard, as it may represent a weak spot in the bottle.

The proper fill levels in bottled beer are about halfway up the neck of the bottle. If the fill level is too low or too high you may experience problems.

58 My bottle caps are rusty

CAUSE

Bottle caps have been stored wet or in a moist environment. Rusty bottle caps can leave a layer of rust on the lip of the bottle, and may not form an appropriate seal for carbonation.

SOLUTION

Under no circumstances should you use rusty bottle caps for bottling beer. If the caps on your bottled beer have rusted, investigate your storage environment for excessive moisture. Be sure to wipe any residual rust from the bottle before drinking. While consuming rust isn't dangerous in and of itself, it can contribute undesirable flavours to your beer.

The main contributor to rusty bottle caps is returning bottle caps to closed storage after sanitising on a bottling day, when more bottle caps have been sanitised than actually used. While it is an excellent idea to have more bottle caps than needed to account for any mistakes, drops or damage that may occur during the bottling process, it's a good idea to fully dry bottle caps before moving them into a closed bag, box or other storage container. It's most efficient to lay them out in a single layer, crown side up, on a paper towel in a dry, warm place.

Be sure to re-sanitise all caps immediately prior to their next use.

59 My bottled beer has too much carbonation

CAUSE

Beer was not fully attenuated prior to bottling, or too much priming sugar has been used, or there is a wild yeast or bacterial infection causing excessive fermentation in the bottle and producing a large amount of CO_2.

SOLUTION

These beers, usually known as 'gushers', can be an unpleasant surprise when opening bottles for the first time. They can also be *very* dangerous.

Most glass bottles are rated three-to-four volumes of CO_2, which is about 60 to 100 per cent more pressure than a normal homebrew should see. However, bottles can have flaws and over time, with frequent reuse, can develop chips and cracks, making them more likely to explode under excessive pressure.

To safely handle over-carbonated bottles, chill them as much as you possibly can without freezing them. CO_2 dissolves into liquid more easily at lower temperatures and the bottles will be safer to open when cold. Consider wearing gloves and eye protection when opening bottles.

If you'd like to recover the beer, very carefully remove the bottle cap, just allowing a hiss of CO_2 to escape slowly. Allow the bottle to warm about 10 degrees – which allows more CO_2 to come out of solution – and recap with new, sanitised caps. There will be a larger than normal amount of carbonic acid present in the beers, which can be recognised as a slightly metallic taste.

To prevent over-carbonation, be sure to follow cleaning and sanitation procedures for bottles, caps and bottling equipment, and pay attention to priming sugar measurements; a little goes a long way. Finally, be patient! Beer that hasn't finished fermenting can have problems, other than over-carbonation in the bottle, that manifest as unpleasant flavours.

60 There is a lot of yeast in the bottom of my bottles

CAUSE

Beer was bottled before yeast dropped out of solution, or yeast was roused as the beer was transferred into the bottling bucket.

SOLUTION

While it's necessary for there to be some yeast in solution for bottle conditioning to occur, it does not have to be visible to the human eye. If fermentation has finished and the beer still appears cloudy, it is worth spending a few extra days to allow the yeast to drop out of solution before bottling.

If fermentation is complete, you've been waiting, and the beer is still cloudy, check to make sure the beer is not infected with bacteria by tasting some. If it tastes fine, consider adding a post-fermentation fining agent and mixing it in. Liquid fining agents, such as gelatin or silicic acid, can aid in clarification by helping yeast cells and proteins stick together in the solution, thereby dropping out through gravity. Just be sure that all of the tools you use are cleaned and sanitised before introduction into the finished beer.

If you are able, transfer via siphon rather than pouring the beer from one vessel to another. Siphons can create a gentle, slow transfer that allows the brewer to leave sediment behind while transferring nothing but the clearest liquid. In addition, it prevents the excessive introduction of oxygen to the beer, thereby avoiding staling compounds that derive from oxidation.

A cloudy bottled beer could be caused by yeast that escaped the fermenter, or it could be signs of a bacterial infection.

A HOME AUTOCLAVE

An autoclave is a device used for sterilisation in many commercial applications. It is a sealed chamber that uses high temperature and pressure to create an environment in which yeast and bacteria cannot live. At home, the resourceful brewer can use a conventional pressure cooker as a home autoclave. They can be particularly useful for sanitising small pieces that are heat/pressure-resistant, such as silicone or stainless steel, or for sanitising plate heat exchangers.

61 My bottles have ropy filaments in them

CAUSE

The beer has a pediococcus infection. Pediococcus is a gram-positive cocci-shaped (spherical) bacteria that can eat residual sugar in the beer, creating lactic acid and diacetyl off-flavours as well as slimy, ropy filaments.

SOLUTION

A pediococcus infection is considered one of the most difficult infections to clean out of a professional brewery, much less a home brewery. Similar to a lactobacillus infection (see Problem 56), it's best to start at the end product, and work backwards to investigate cleanliness at each step of the process.

However, to be safest, you should simply replace any soft, scratchable or porous material that you can't effectively soak or access the inside of. Hoses, bottling wands, any gaskets in your operation, even a plastic racking cane may be the source of a potential infection.

Take all remaining pieces apart and thoroughly clean each (bear in mind that plate heat exchangers can be incredibly difficult to put back together!), scrubbing each part. Pay close attention to any cracks, crevices or seams, and scrub closely to disrupt any possible biofilm that may have formed. Use a caustic cleaner on all parts. Be sure that the caustic is as hot as you can get it, then follow this with a cold sanitation rinse. Allow each part to completely air-dry. Sanitise again before use.

Finally, review your cleaning and sanitation procedures before your next brew to be sure you avoid this infection in the future.

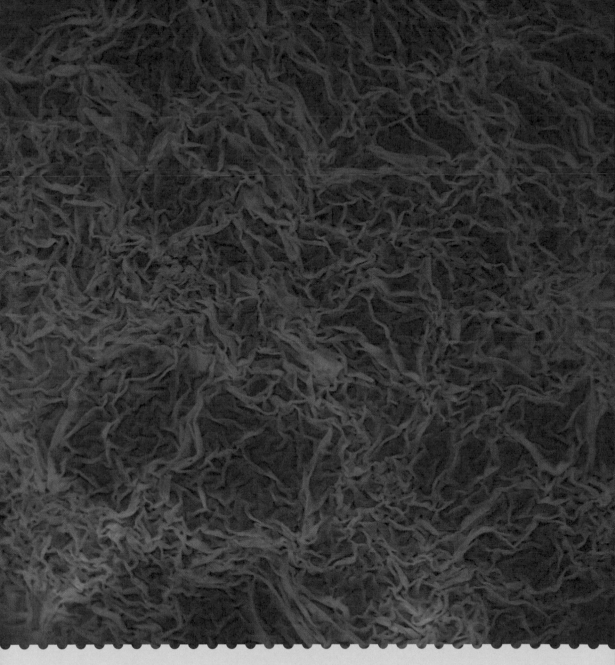

Pellicles are signs of wild yeast or bacterial infections in your beer and shouldn't be taken lightly.

62 There's a ring around the beer line in my bottle.

CAUSE

The beer has an infection of bacteria or wild yeast or, rarely, there is a mini 'krausen' ring from re-fermentation in the bottle that sometimes results from using dry malt extract or some unrefined sugars.

SOLUTION

A ring around the inside of the bottle is almost definitely a sign of some sort of infection. A white film across the top of the beer in the neck of the bottle is a sure sign that the ring derives from an infection rather than anything else and should be treated as such. Wild yeast is often a little less tricky to remove than bacteria – after all, you remove yeast from your equipment on a regular basis – but without knowing the exact kind or source, it's worth treating every infection as a worst possible scenario (see Problems 56 and 61).

In some rare cases, the re-fermentation used for bottle carbonation can create a mini krausen ring inside the neck of the bottle. Exercise caution if you think this may be the case, as a fermentation robust enough to leave a krausen ring is also likely to create a significant amount of CO_2 within the package and may lead to a bottle that is under an extraordinary amount of pressure. There was almost definitely too much priming sugar, too much yeast remaining in the solution, or both!

In either case, take extra care when cleaning these bottles for reuse. Any residual cells from a ring inside the neck of the bottle, whether it is from a bacterial infection or yeast krausen, can affect your next batch of beer. Scrub the inside of each bottle with a bottlebrush before soaking, or consider replacing the bottles altogether.

A pellicle can spread even more inside a bottle if wild yeast or bacteria is present and can be an easily identifiable sign of infection.

63 My mini-cask or pressure barrel is over-conditioned

CAUSE

Over-conditioning can cause your mini-cask or pressure barrel to fob when pouring, or pour slowly. Priming cask-conditioned beer requires much less priming solution than bottle conditioning.

SOLUTION

Home-brewers who wish to dispense naturally conditioned beer at home have several options open to them, the most common of which is bottle conditioning. It's also possible to use five-litre mini-casks or pressure barrels to serve real ale at home. Mini-casks are sealed containers which you prime in a similar way to bottle conditioning, however they require much less sugar or other priming agent in order to reach the desired level of carbonation.

A mini-cask will condition to a good level with just 10 g of sugar and will be ready to serve within 1–2 weeks if the cask is left at fermentation temperature. The casks feature a top seal that can be used to vent off any excess carbonation. Once the cask has been breached, leave this vent slightly open to ensure a good, steady pour and if required, adjust the pouring rate by rotating the tap slightly.

Pressure barrels feature a pressure release valve which should release any excess carbonation automatically. If the barrel is seen to be bulging, you can unscrew the cap slightly to release the excess pressure and vent the beer. This can be caused by the cap becoming blocked, especially if the barrel was used for primary fermentation. Once the beer is ready to serve, as you start to dispense the beer you'll need to either let air into the top of the barrel to ensure good flow or apply top pressure using a CO_2 canister.

THE FLAVOUR OF CO₂

CO_2 has a distinctive flavour. It is the flavour of carbonic acid, and nowhere is it easier to taste than in over-carbonated beers. People often refer to over-carbonated beers as having a 'bite', but beyond the prickly sensation on the tongue from too much carbonation, CO_2 has a slightly acrid, bitter flavour that can persist in a beer, even after the bubbles have broken out and gone away.

Over-carbonation makes beer tricky to pour without producing excess foam. Use a cool, clean glass held at a 45-degree angle to the tap and pour the beer slowly down the side of the glass, straightening the glass as it fills.

64 Beer poured from my pressure barrel is flat

CAUSE

Secondary fermentation in a pressure barrel produces the CO_2 required to naturally condition the beer. If that CO_2 escapes, the beer will go flat.

SOLUTION

Beer fermented in or transferred to pressure barrels after fermentation can be primed to allow it to condition naturally as real ale. The secondary fermentation will produce CO_2 that remains in the beer to provide the desired level of carbonation. However, a number of potential problems exist that might prevent the beer carbonating correctly.

Pressure barrels are fitted with pressure relief caps, designed to allow the barrel to vent off CO_2 to avoid a build up of excess gas within the barrel which can cause over-carbonation. The most common problem is that the cap is not sealed correctly against the top of the barrel. If you suspect a poor seal, unscrew the cap from the barrel and remove the seal from inside the cap. You can then drop the seal in a cup of hot water to help it go back to its original shape before cooling it in a cup of cold water and replacing it in the clean and sanitised cap. A thin film of vaseline along the surface of the seal can help it form a good bond with the barrel, but be careful not to use too much as this can cause the seal to slip within the cap. To replace the cap make sure it is cooled after cleaning and screw it down until the seal touches the neck of the barrel before turning it to tighten – a quarter turn should be sufficient.

If you're still having problems after ensuring a good seal, consider replacing the cap.

Beer can carbonate naturally, or it can be force-carbonated, depending on the style of beer that is being made.

65 I need to force-carbonate my keg really quickly

CAUSE

Beer has finished fermenting, but too late to take the time to properly carbonate before the event you'd like to serve it at.

SOLUTION

Getting beer to carbonate is a function of temperature, pressure and time. In many home-kegging environments, carbonation is achieved by pushing CO_2 into the head space of a keg and waiting for the carbonation to fully dissolve into the liquid, which can take days. To achieve carbonation more quickly, you might try the following:

- Get a carbonation stone. This is a small, porous ceramic, or stainless steel piece that is put inside your keg, through which you can inject CO_2. Small CO_2 bubbles dissolve into liquid faster than large ones. The colder the liquid is, the more efficiently CO_2 will dissolve into solution.

- Lie the keg on its side while carbonating. This increases the amount of surface area available for CO_2 to dissolve into liquid. If you haven't cleaned and sanitised every internal surface of your keg well, this may lead to an increased infection risk.

- Shake the keg. Shaking it while under pressure will allow some of the CO_2 to dissolve into liquid quickly. This allows more CO_2 to enter the head space of the keg, keeping there what's already in the solution.

- Increase the pressure. CO_2 pushed at 30 psi at 3°C (38°F) will carbonate a keg much more quickly than 14 psi at 3°C. However, there is a risk of over-carbonation if you're not careful.

- Combine several of these methods. Placing a 3°C keg on its side, with CO_2 at 30 psi, while occasionally shaking the keg, can carbonate beer in a matter of hours. Again, be careful of over-carbonating while using this method.

CO₂ CHARTS AND VOLUMES

CO_2 charts show the correlation between temperature and pressure, expressed in volumes of CO_2. A single volume equals 1 litre of CO_2 at 20°C (68°F) at 1 atmosphere (14.7 psi) in 1 litre of beer. Most beer is carbonated to between 2.2 and 2.6 volumes of CO_2 but some can be much lower (1.5 for cask ales) or much higher (3–5 volumes in wheats or some Belgian beers).

🍂 *Pouring a beer from a well-balanced draft system should take about eight seconds, finishing with about a half-inch collar of foam at the top of the glass.*

66 My Cornelius keg won't seal properly/is leaking

CAUSE

Gaskets are worn or cracked, springs in posts are worn, or the lid of the keg is not sealed well.

SOLUTION

It's important to be comfortable with taking all parts of a Cornelius keg apart and putting them back together again, as it is the only way you can be sure that it is perfectly cleaned and sanitised. It is also vital for identifying gas leaks during regular use. Posts unscrew from the keg. Beneath each post is a dip tube: one short one, which barely enters the keg (the gas tube), one long one that reaches to the bottom of the keg (the liquid tube). The dip tubes have identical gaskets that allow the post to seal with the keg. Each of these gaskets should be free of any cracks, and should snugly fit around their dip tube.

Each post has a poppet valve inside which consists of a spring and a poppet gasket. The spring should easily keep the poppet gasket pressed up against the top of the post. These poppets are entirely replaceable.

Each post also has an O-ring on the exterior of the post that allows fittings to attach snugly and securely. These O-rings should fit into a small groove on the post and have no cracks or nicks in them.

The posts are interchangeable on pin-lock kegs but not on ball-lock kegs, which have separate gas and liquid-style posts. Use silicone tape on threads to create an airtight seal.

The lid has a large O-ring that should be easy to slip on and off. The O-ring should not have any cracks or nicks and should not be worn in any spot. On used kegs, the opening where the lid fits may be slightly warped and may have difficulty sealing without pressure. Pulling up on the O-ring while pressuring the keg may sometimes help create a seal. You may try coating the O-ring with a food-grade silicone grease.

FORCED VS. NATURAL CARBONATION

A Cornelius keg introduces a specific level of carbonation by adding top-pressure to the beer, which over time absorbs the CO_2. Once carbonated, pressure is reduced to a level suitable to dispense. To naturally condition real ale in a Cornelius keg, prime the beer as you would when bottling, seal it by briefly applying pressure, then store at fermentation temperature. Vent the keg to remove any over-carbonation. Once carbonated, dispense as normal.

Cornelius kegs are made of stainless steel, which makes them pretty hardy. But repeated use over time can mean intrinsic parts get worn out and need replacing.

67 My kegged beer gets over-carbonated/flat

CAUSE

The pressure that beer is being served at in the kegerator is set too high or too low.

SOLUTION

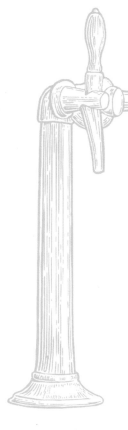

Serving beer correctly is the act of pushing enough CO_2 into a keg to keep CO_2 in solution, while also pushing it up a hose and out into a glass. When pouring a glass of beer, liquid should be flowing at the rate of roughly 2 ounces per second without excessive foaming.

The most common set up for home kegerators is a 5-foot-long (1.5 m) vinyl hose with a 3/16 inch (4.8 mm) internal diameter leading from the keg to the tap. In such applications, depending on the carbonation level in the beer, the gauge on the CO_2 tank should be set between 12 and 14 psi for a proper pour. If the pressure is set too low, the beer will de-gas as it is being poured. If the pressure is set too high, the amount of CO_2 dissolved in the beer will increase over time. If your kegerator is colder than 3°C (38°F), reduce the pressure on the CO_2 tank gauge. If your kegerator is warmer than 3°C, increase the pressure. Consult a CO_2 chart for pressures that correspond with ideal levels of CO_2. You shouldn't need to adjust more than 1–2 psi for temperature.

If your hoses are longer, of a different material, or of a different diameter, consult the manufacturer's information about the amount of resistance present in each foot of hose and adjust accordingly. Different materials can vary widely. A useful website for specific information about a wide variety of materials and draft set-ups is www.draughtquality.org.

If your kegged beer shows uncontrollable carbonation there might be a fault with the pressure or the temperature in your keg.

68 My kegged beer tastes like rancid butter

CAUSE

The draft lines leading from the keg to the tap are dirty. Draft line infections are usually due to pediococcus growth, which can spread down the line and into the keg, infecting the entire batch of beer if not properly dealt with.

SOLUTION

In a commercial setting, draft lines should be cleaned every two weeks. While that may seem impractical in a home setting, maintaining clean draft lines can mean better beer and less waste while you're pouring homebrew, so it's important to keep up with it.

While you can purchase specific beer line cleaner, you can use the same caustic cleaner and sanitiser that you use to clean and sanitise your brewing equipment in your draft lines.

Begin by draining the lines and flushing them through with hot caustic solution. Ideally, you should allow the hot caustic solution to rest in your beer lines for at least 20 minutes before flushing out with clean water. Disassemble and clean the taps and any fittings which touch beer when you clean the beer lines. If any of those pieces have infections and are not cleaned, they can spread back to the beer lines quickly and easily.

Flush cold water through the lines until there is no visible debris being flushed from the lines. If you have access to pH strips, use pH strips to measure the pH of the liquid to be sure that it matches tap water.

If possible, repeat the process every three months with an acid cleaner to help remove beer stone. Beer stone is calcium oxylate – a calcium that can precipitate out of beer and settle into your beer lines, causing foaming, and giving bacteria microscopic cracks and crevices to live in. Beer lines should be replaced entirely every year.

❧ *Even the best made beer is only as good as the method in which it's stored and served. Lines should be cleaned as often as possible, and replaced entirely once per year.*

69 I don't know how to bottle beer from a keg

CAUSE

Whether bringing homebrew to a friend's house or entering beer into a competition, there are times when packaging kegged beer into a bottle is desirable, but it's not always easy to figure out how to do this.

SOLUTION

In a pinch, you can just fill a bottle from the tap and cap it. The shelf life of that bottle might not be the greatest – you'll want to consume it within a few days – but it will work.

The best possible scenario is to clean and sanitise a bottle, purge it with CO_2 and then fill it under pressure using a counter-pressure bottle filler, or a beer gun. These are available from your local homebrew supplier and are easy to operate. The idea is that the bottle is pressurised to the same pressure that the keg is, then the beer is filled into the pressurised environment, releasing CO^2 from the bottle at the same rate that beer enters the bottle, reducing foaming and loss. When the bottle is filled, the filler is removed, a cap is placed on the foam coming out of the bottle, and the remaining CO_2 is trapped inside a clean, sterile, pressurised environment.

If a counter-pressure filler is unavailable, be very sure that the bottles are cleaned and sanitised, and, if you can, find a way to purge oxygen from the bottle using CO_2 before filling.

Finally, if you can find a hose that will fit over the end of your tap, and also reach to the bottom of a bottle, you can fill from the bottom up – pushing air out of the bottle, rather than splashing beer into the bottle – to minimise the bottle's oxygen content.

A beer gun is used to fill bottles with beer that has already been carbonated and in a keg. Beer guns are useful for making small amounts of beer portable or to enter them into competitions.

CHAPTER SEVEN
EQUIPMENT

Equipment is the largest variable in home-brewing. Beer can be made at home using the simplest, most rudimentary tools that get the job done. It is a matter of choice whether to use DIY pieces hacked together in your garage, or a more sophisticated set-up that looks and functions like a scaled-down model of a professional brewery.

Equipment problems often stem from pieces of equipment breaking or not functioning as they're supposed to. No matter how expensive or sophisticated your set-up is, the most common cause of problems is neglecting to properly clean and sanitise the equipment. Cleanliness is paramount and necessary to making great beer at home.

Often, the more complicated equipment comes with a host of new problems, such as new dead spots, creases, cracks and crevices – barely visible to the casual observer – where mould or bacteria can grow and thrive, infecting your batches.

Maintaining equipment on a regular basis and becoming intimately familiar with its construction can make a brew day – and making great beer – considerably easier.

70 I don't have a big enough pot to brew with

CAUSE

Sometimes because of space or cost considerations, a brewer has smaller equipment than the batch size that they'd like to make.

SOLUTION

It is possible to make a concentrated, high-gravity wort and dilute it with fresh water to appropriate volume in the fermentation vessel. This process is known among home-brewers as a 'partial boil'.

The process is easiest to do when brewing with extract, but is also available to all-grain brewers, and in many ways is simpler than brewing a full boil, because it removes the need for rigorous and frequent water measurements. Add the extract as you normally would, but in a smaller amount of water. Then add water at the end of the process, as you would with any other brew.

All-grain brewers will see a drop in efficiency when running off less wort than normal, and may need to increase the amount of grain in the grist accordingly. Unfortunately, with the wide range of available equipment, it's almost impossible to predict how much more grain would be needed.

Finally, there are a couple of other things to consider:

- Maillard reactions, such as caramelisation, are more pronounced in high-gravity worts. If using extract, try using the lightest possible extract you can. All-grain brewers can adjust recipes a little more easily, but in both extract and all-grain cases, some lighter styles may be slightly out of reach using this method.

- Hops utilisation and isomerisation – the amount of bitterness derived from boiling hops – is reduced in higher gravity wort. In addition, the more you dilute your wort, the more you dilute your bitterness.

Ensuring your pot is the correct size can make a real difference on brew day.

71 I can't get a siphon started

CAUSE

Siphons are the ideal way to transfer liquid from one vessel to another when home-brewing, and it seems easy, but a good siphon can be tricky to achieve.

SOLUTION

Siphons are more difficult to start when there is not a significant drop in height, or with fluids with higher viscosity (like wort) – or worse, with both!

To start a siphon it helps to have to hand-distilled water or sanitising solution. Fill the hose and racking cane with the liquid and then pinch off the end of the hose to stop the liquid from draining out. Some amount of liquid may drip out of the end of the racking cane into your wort, so be sure that the liquid that you're using can come in contact with your wort and/or is free of any possible contaminate.

Once the racking cane is in place, position the end of the hose below the bottom of the kettle and allow the liquid to drain out. This should start a siphon. Once the sanitary liquid has been cleared from the racking cane and hose, wort (or beer) should be flowing and you can safely transfer the hose into the intended vessel.

You may also seek out a specialsed piece of equipment that is a self-starting siphon/racking cane. This is built as a double-walled tube. The exterior tube fills with liquid, while the interior tube is a racking cane. A siphon is started by operating the two together like a pump. Once liquid is flowing, a siphon is started. However, these pieces of equipment need extra care to clean, as there are many small parts that are submerged in wort or beer. They can be very easy to use, but can lead to infection if not cared for properly.

GO FOR THE MOUTH

The easiest way to start a siphon is to submerge one end of a hose in liquid and then suck on the free end until liquid is flowing, while making sure that the liquid is flowing from a high point to a low one. Gravity will do all the work, so proceed with caution: a mouth siphon might be easy, but it's a possible source of contamination and a dangerous way to siphon chemicals or boiling liquid!

An auto-siphon, as seen here, is a simple piece of equipment that can make it quick and easy to get a siphon started.

72 My hoses get foggy after sanitising

CAUSE

The walls of some clear, vinyl hoses will become cloudy when in the presence of some acids, particularly if soaking for a long period.

SOLUTION

While it can be worrying to notice that hoses that were once clear are now cloudy, they should still be okay to use for brewing, assuming you properly clean and sanitise the hoses before use.

Some foaming acids used for cleaning may cause vinyl to become slimy after long-term contact, and while this may seem alarming at first, it is merely a chemical breakdown of the acid. Foaming acids contain what is called a 'surfactant', a chemical compound that lowers the surface tension between a liquid and a solid, which allows the acid to have lingering contact time with the material it is sanitising. Over time, that surfactant may separate from the solution, particularly after the acid has been diluted for use, leaving a slippery feeling.

Most sanitisers used in home-brewing do not require soaking, but are quick-action sanitisers that only need a few minutes of contact time to do their job. Rather than soaking items in a sanitiser bath, consider filling a spray bottle with sanitiser for spraying on and in the pieces that need to be sanitised.

Many sanitisers lose their efficiency over time after being diluted and/or exposed to oxygen. Sanitiser can neutralise over time, leaving an excellent substrate from which bacteria can grow. Always leave your hoses and fittings dry in between uses.

When in doubt, hoses are relatively cheap to replace. Hoses should be regularly replaced if they're discoloured, cracked or otherwise damaged.

❧ Brethren hoses can be soaked inside a sanitising bath in a demijohn, as seen here. Alternatively, if you are using a quick-action sanitiser, avoid soaking all together by adding the solution to a spray bottle and using it that way.

SILICONE VS. VINYL

If you can afford to, it's often worth spending the extra money on silicone hoses. They are durable, heat-resistant, and can withstand caustics, acids or even steam. You can even put them into an autoclave for sterilisation.

73 My mash efficiency is low

CAUSE

Low efficiency is most commonly caused by not crushing grain finely enough, but there are other actions a brewer can take to help improve efficiency.

SOLUTION

First and foremost, be sure to mill grain finely. A good crush should have no whole barley kernels in it; the mill should have cracked open all the kernels but left the hull of the barley intact. Crushing grain too finely can lead to increased tannin extraction from the hull, or a stuck mash (see Problem 26), but does mean increased efficiency.

You may consider mashing for a longer time to allow more time for starches to convert into sugar (but check Problems 28 and 29 for potential problems with this option).

Be sure that you're sparging with hot water. Water at 76°C–77°C (168°F–170°F) can extract sugar from grain more efficiently than at cooler temperatures. But it will also arrest the enzymatic reactions. Be sure that your mash is complete before sparging. If you sparge with water that is too hot, unpleasantly bitter tannins can be extracted from the barley hulls.

When you do sparge, slow down the rate of flow. It can take time for sugar to solubilise into water. If your sparging is too fast, you may be leaving sugar behind in the grain.

Finally, be sure that you minimise the losses in your brewhouse. Extra-long hoses, frequent spills and splashes, or dead spaces in the mash tun can leave wort behind, and in that wort is valuable sugar. An important part of brewhouse efficiency is collecting all the wort that you can possibly generate.

TESTING FOR STARCH

If you're not sure you're mashing long enough to convert all of the available starch into sugar, do an iodine test. To do this, pull out a small portion of your mash and add a drop of iodine into it. Iodine turns blue in the presence of starch. If there's still starch in the mash, consider allowing more time for starch conversion to finish.

Stirring the mash (before lautering begins) can help starch conversion and help ensure that you don't have any clumps of dry grain interspersed throughout the mash bed.

74 My boil kettle has brown or white build-up in it

CAUSE

Residual proteins, beer stone or scale have settled out into the kettle. These can create nucleation points in the kettle and cause boil-overs, and sometimes lend undesirable flavour or colour characteristics to the final beer.

SOLUTION

It's important to realise that there are two unrelated issues that create this problem, and sometimes these issues need to be addressed separately.

A white build-up on the sides of your kettle can mean that calcium is settling out of the water as you boil and sticking to the sides of the kettle. This is not a problem by itself; in fact, it's exactly what should happen over time. If calcium is falling out in the kettle, it is not getting into your fermenter, bottle, keg or draft line.

A brown build-up is often the result of brewing with high-protein grain. Protein can be gummy and can stick to the sides of the kettle – particularly if there is an existing calcium build-up.

Organics (protein) are generally removed with caustic cleaners – be sure to use a large dose of caustic at a hot temperature (60°C/140°F or higher) and soak the entire kettle for at least 30 minutes. After the caustic is removed, the kettle may still have build-up. Scrub the sides of the kettle with a soft scrub pad to remove this, then rinse with cold water.

Removing calcium build-ups may require an oxidative caustic, or an acid. Check your local homebrew shop for acid cleaners. In some cases, soaking the kettle in white vinegar may break down and remove some of the beer stone, but it may not be enough to help large build-ups.

❧ Cleaning is one of the most important tasks in brewing, and often nothing does the trick better than diving in and scrubbing by hand. Be sure to use the soft side of your scrub pad to avoid leaving scratches that can habour bacteria.

75 The flow through my plate chiller is too slow

CAUSE

The inside of the plate chiller is dirty, causing flow to be restricted. This is also a potential infection hazard.

SOLUTION

Plate chillers can be very useful tools for helping cool wort at an incredibly fast rate, but they are also one of the only pieces of equipment that a brewer cannot see inside. It's hard to tell when the inside of a plate chiller is getting dirty. The only ways to find out are by monitoring flow rate or by tracing an infection back to the plate chiller.

A word of caution: only take your plate chiller apart if you absolutely feel it is necessary. They are notoriously difficult to reassemble, and you may find you have created far more problems than you've fixed.

Cleaning a plate chiller is best done by using a pump and an oxidative caustic cleaner. Run hot caustic through the plate chiller for at least 15 minutes. If possible, restrict the flow on the exit port to be sure that caustic is completely filling the plate chiller. Rinse by doing the same with cold water until a pH strip inserted into the exit water reads neutral.

If you do not own a pump, completely submerge the plate chiller in the cleaning solution. Wear gloves and turn the plate chiller over within the liquid repeatedly to be sure that all air has escaped, and that the cleaning solution is in contact with all surfaces. Soak for at least 15 minutes, and use a hose to force water through at a high pressure to dislodge any solids that may remain.

✿ *Pumps can cause bubbles in a hose line, which can present problems. Bubbles can mean that air is getting mixed in somewhere, which could oxygenate wort or cause an infection.*

PRIMING THE PUMP

When working with a plate chiller, it's often best to use a pump. Very few of the pumps that are sized for homebrew operations are self-priming pumps. The pump needs to be filled with liquid before flow will start. Be sure to use gravity to completely fill the pump before turning it on, or fill with sterile water before adding beer.

76 I don't know how to sanitise my plate chiller

CAUSE

Plate chillers may be clean, but it can be difficult to know if sanitising solution has touched and soaked every portion of the inside of a plate chiller.

SOLUTION

Sanitising cleaned equipment before brewing is vital for making good homebrew. A plate chiller is the first item that your finished wort touches, and so the first place where it is vulnerable to infection. While it is good practice to soak your plate chiller in sanitising solution, or pump sanitiser through it, it is incredibly difficult to know whether every bit of the inside of the plate chiller has been properly sanitised.

First, be sure you know how to remove all rubber parts from the plate chiller. These are normally just gaskets on the fittings for beer and water flow. Removing these rubber gaskets gives you a wider range of options for sanitising your plate chiller.

A home pressure cooker can act as an autoclave for stainless steel and silicone parts, and can be ideal for sanitising a plate chiller. If a pressure cooker is not available, you can sanitise your plate chiller by adding it to an oven at 177°C (350°F) for 30 minutes. Before putting a plate chiller into either a pressure cooker or an oven, cover the open ends of the chiller with tin foil. When the chiller is removed from either of these environments, it will be very hot. Handle it with care and allow to cool before replacing rubber gaskets and using it for home-brewing. The foil will protect the openings while the plate chiller is cooling.

A home-brewer can make great use of a domestic pressure cooker for sterilising equipment; it can function well as a home autoclave.

77 I'm not sure how much liquid is in my kettle

CAUSE

Homebrew recipes often contain very specific volume measurements, but very few pots or vessels large enough to boil the requisite amount of liquid are marked for volume measurements.

SOLUTION

The easiest way to solve this problem involves planning ahead. You need just a few simple materials. You'll need a large marked measuring jug and a wooden or plastic rod (you could use the spoon or ladle you use for brewing). Simply add a measured amount of water into your brew kettle, then dip your spoon or rod into the liquid and mark off the level. Repeat this process at increment levels (e.g. quarts, gallons, litres). You now have your own measuring stick. On the brew day, insert this into the liquid in the kettle to check your volume.

If you're already in the middle of a brew day you can determine the volume of your kettle with a simple formula: *Volume*= π $r^2 h$ (r = the radius – or half the diameter – of your kettle, h = the height of your kettle on the inside).

For example, if your kettle is 1 foot wide and 2 feet deep, the volume of your kettle would equal $\pi \times 0.5^2 \times 3 = 2.36$ cubic feet. Since one cubic foot is 7.48 gallons, your pot can hold 7.38×2.36 gallons (17.41 gallons).

To determine the amount of liquid per inch of depth, simply divide 17.41 by 24 inches. In metric 1 cubic cm = 1 ml. A kettle 30 cm wide by 50 cm tall contains 35,343 cubic cm, or 35,343 ml (35.343 l); 1 litre per 7 cm of height.

🌿 *A simple handmade measuring device can be a valuable piece of kit for homebrewers. Measuring devices can take the form of a spoon that you use to brew with or a simple marked rod. It needs to be at least the full height of your kettle.*

78 I'm not sure if my mash tun is big enough

CAUSE

Homebrew recipes can take widely differing amounts of grain and can stretch the limits of small mash tuns. It can be difficult to know if a recipe will fit in your equipment.

SOLUTION

One of the easiest ways of finding out how much liquid will fit in your mash tun is simply trial and error. Unlike finding the volume of your kettle, it can be a frustrating process, because you are potentially wasting expensive ingredients if your mash tun won't fit the recipe you're attempting to make.

You can easily find the volume of your mash tun using the method described in Problem 77:

One pound of crushed grain will take up roughly 295 ml (10 fl oz) of space while dry. Typically home-brewers use a water-to-grain ratio of 1.25 quarts (40 fl oz/1.18 l) per 1 pound (454 g) of grain, which means that for each pound of grain in a recipe you can expect to use 50 fl oz of space, or roughly 1.5 litres (0.4 gallons). If you are using a 20-litre (5-gallon) cooler for a mash tun, you can expect to be able to fit 5.5 kg (12.5 lb) of grain, plus the water needed to make a batch.

Varying the amount of liquid can allow room for more grain, but not as much as it might seem. Reducing the water-to-grain ratio to 1 litre per 453 g (1 quart per pound) means that each pound of grain, plus water, takes up 1.25 litres (0.33 gallons) of space, allowing you to fit a little over 6.8 kg (15 lb) of grain into a 20-litre (5-gallon) mash tun. Not a substantial difference, but it could allow you the required extra space.

❧ *The amount of grain that you can fit into your mash tun dictates how high the alcohol of your beer can be.*

79 I don't have a way to oxygenate my wort

CAUSE

Brewers often oxygenate wort prior to adding yeast to create an oxygen-rich environment for yeast to grow in, but without specialised equipment, including an oxygen tank and an oxygen stone, this can be a daunting task.

SOLUTION

Yeast need oxygen to grow strong cell walls. Each strain of yeast has its own unique oxygen needs. It's safe to assume that the harsher the environment yeasts are introduced into, the more oxygen they need. Higher-gravity beers, beers with a high hop content, or even the cool fermenting temperatures of lager yeast strains, all require a bit more oxygen.

Fortunately, introducing oxygen at home is easier than it sounds. For most beers, operating in the open environment of your home brewery, splashing wort down into your fermenter or bottling bucket, will introduce enough oxygen for most yeast strains to take up what they need to grow.

When in doubt, shaking the fermentation vessel rigorously after it's filled can introduce extra oxygen into solution, and can help yeasts get going in tougher situations. Take care if doing this, because the fermentation vessel can be very heavy.

Some studies suggest that the use of oleic acid can, in place of oxygen, sometimes help yeast build up cell walls. The most common source of oleic acid is olive oil, but dosing rates are so low that it is almost impractical to use at home. Dipping a toothpick into olive oil and using the drop that forms on that toothpick is more than enough for a 20-litre (5-gallon) batch.

❧ *Pouring wort into a fermentation bucket is the easiest and most economical way for aerating wort at home, but this process can lead to trub and spent hops from the bottom of your kettle being transferred along with your wort.*

80 My CO₂ cylinder empties too quickly

CAUSE

A full 2 kg (5 lb) tank of gas should last a long time in a homebrew environment, but leaks in connections can turn a CO_2 cylinder into a costly accessory.

SOLUTION

A 2 kg (5 lb) CO_2 cylinder should be able to serve about fifteen 20-litre (5-gallon) kegs of homebrew (roughly 155 g/ 5.5 oz of CO_2 per keg). You'll need about 100 g (3.5 oz) of CO_2 to carbonate a 20-litre keg to a standard carbonation level. For every keg carbonated and then served, you'll use about 225 g (0.5 lb) of CO_2 in total, or about 10 kegs per 2 kg (5 lb) tank. This assumes that CO_2 is not being used for any other tasks in the home brewery, that draft lines are balanced correctly, that each keg is filled to an equal volume, and that the CO_2 cylinder has exactly 2 kg (5 lb) of CO_2. Obviously, usage may vary slightly.

If usage is considerably lower than 10 kegs per 2 kg (5 lb) tank, then you may have a leak. First, fill a spray bottle with tap water and a few drops of dish detergent, then spray at every point where there is a connection that gas is travelling through, going from the threaded fitting for the regulator on the CO_2 tank all the way to the keg. If you see bubbles, you've found a leak.

All threaded connections should be sealed with silicone pipe tape to be sure that no gas is passing through the threads. For barbed hose connections, be sure that hose clamps are used at every juncture and tightened regularly. It may take time and attention to find and resolve CO_2 leaks, but the cost savings can be worth it.

TYPES OF CLAMP

If after tightening the clamps you still have leaks, consider changing the type of clamp being used. For most applications, jubilee style clips with adjustable screws should be sufficient to provide an air-tight seal and avoid leaks. Crimped clamps work slightly differently in that a special tool is used to tighten the clamp around the hose. These may provide a tighter seal but can only be used once and can't be easily adjusted once in place.

A CO_2 cylinder should be handled with care and stored securely when not in use in order to avoid damage that may lead to leaks. Every hose junction and every place that requires a hose clamp is the location for a possible leak. Be sure to tighten these regularly and check them for leaks.

81 My airlock blew off my fermenting vessel

CAUSE

Pressure built up inside a fermenter caused the airlock to dislodge and pop off, exposing the beer inside the fermenter to possible contamination.

SOLUTION

While this is often the result of an overflowed fermentation, in a case where the airlock doesn't fit snugly in the top of the fermenter, this can also result from just normal pressure. In any case, it can be alarming to find out that fermentation is not protected.

Fortunately, fermentation protects itself to some extent. The pressure in the fermenter that blew the airlock off is often enough to keep other microbes and debris out of the fermenting liquid. As yeast is digesting sugar, it expels CO_2, which is heavier than oxygen, and will create a blanket on the top of the fermenting liquid, preventing oxygen and other airborne bacteria or yeasts from reaching the surface.

However, this will not stop large debris or flying insects from entering the fermentation. Fruit flies are particularly attracted by the scent of fermentation and the presence of CO_2. If fermentation is vigorous, krausen at the top of the beer should act as a protective barrier. Take time to visually inspect the top of the beer, and remove any foreign debris before sealing the fermentation and replacing the airlock.

In most cases, the beer will be perfectly safe, and you may continue as normal, but be sure to check flavours and aromas as fermentation progresses for any potential signs of contamination.

❧ *A vigorous fermentation can often blow right through an airlock, causing it to pop off. This can cause alarm.*

CHAPTER EIGHT
FINISHED BEER

The reward for all the diligence and hard work is a good finished product. So there is nothing more disappointing than opening a beer, after a long brew day, two weeks of fermentation, a bottling day and two weeks of bottle conditioning, only to find it's sub-par in some way.

It will happen to every brewer at some point, no matter how experienced or skilled. The most important thing a brewer can do is to learn from the experience, put it behind them, and move on to make better beer next time.

Because of this, it can sometimes be advantageous to become closely familiar with a beer that has gone bad, in order to figure out what went wrong. Was it your ingredients? Your process? Something that went wrong during the brew day? A cleanliness issue? Finding out how a particular unpleasant flavour is created can often be the key to preventing it from occurring in the future.

Prevention is the best cure. Once the beer is finished, it's too late to salvage it, but there's still plenty of time to fix every future batch.

82 My beer smells/tastes like butter/butterscotch

CAUSE

The presence of a buttery or butterscotch-like flavour is due to the presence of diacetyl, which is either due to finishing fermentation too quickly, or bacterial contamination.

SOLUTION

Diacetyl is the chemical that is used to give cinema popcorn butter its flavour and aroma. In some beers, it can fit right in, but in many the buttery flavour and oily slickness that comes with it can be off-putting.

Diacetyl is a natural by-product of yeast in fermentation. It is formed by yeast, as part of its normal metabolic life cycle. If you taste any beer halfway through fermentation, odds are there will be an overwhelmingly buttery characteristic to it. However, as yeast runs out of resources to digest in wort, it will turn to diacetyl, digesting this instead, and thereby in the process cleaning up after itself.

If a beer contains a large amount of diacetyl, it is usually because the yeast dropped out of suspension before fermentation had completely finished. Put another way, the yeast went dormant before it could finish eating. This can happen with highly flocculent yeast strains, particularly some English yeast strains. It can also happen if the beer is fermenting in a cool environment.

Finally, the presence of diacetyl may be indicative of a bacterial infection. Pediococcus, as part of its metabolic process, is known to create diacetyl and lactic acid. If, as well as the buttery flavour of diacetyl, the beer features a slight sour tang, or appears to have ropy filaments in it, there is likely a pediococcus infection.

Yeast and hops rise to the surface of a fermentation. Ending this fermentation too early could lead to an off-flavour in the beer.

83 My beer smells/tastes like cabbage

CAUSE

An off-flavour of cabbage or cooked vegetables is caused by DMS (dimethyl sulphide), which is present in the beer due to either bacterial contamination or not properly venting the boil.

SOLUTION

DMS is present in all beer to some extent. The precursor of DMS, SMM (S-methylmethionine), is formed during the germination and kilning of barley malt. As wort is heated above 80°C (176°F), SMM is broken down into DMS. Unless the DMS is subsequently removed from the wort, it will remain throughout fermentation and end up in the finished product.

Fortunately, DMS is relatively easy to get rid of. The boiling point of DMS is only 37°C (99°F) and it will become volatile during a rigorous boil of at least 100°C (212°F). Most homebrew recipes specify a 60-minute boil, though many professional brewers boil for 90 minutes to release DMS from the wort altogether.

It is also important that steam is not trapped in a way that allows it to condense into liquid that falls back into the boiling wort. A lid or a chimney stack on a boil kettle may reduce evaporation rate and water loss, but it can also collect DMS, which drips back into the kettle, making removal of this off-flavour more difficult. Most homebrew that contains DMS is due to steam not being vented properly during a boil.

DMS can occasionally be due to an infection of certain strains of enterobacter from poor cleaning or sanitation of a fermentation vessel. Be sure that fermenters are both cleaned and sanitised immediately prior to the introduction of wort (see Problems 8 and 9), that your yeast is healthy, and you have enough to start a quick, vigorous fermentation (see Problems 20 and 21).

🌿 *A 'cooked vegetable' smell in a finished beer is caused by the presence of dimethyl sulphide (DMS). To avoid this from happening it is essential for wort to be kept on a vigorous rolling boil for 90 minutes.*

84 My beer smells/tastes lightstruck/skunky

CAUSE

Light causes a photochemical reaction in beer containing isomerized hops found in beer. This creates an aroma compound that is very like that found in skunk urine.

SOLUTION

Beer can become lightstruck in a manner of minutes. The culprit is UV-B, which is present in sunlight and many fluorescent lights. This process happens most frequently and most quickly with light-coloured beers that contain more than a lot of hops. Darker beers or hazy beers will not become lightstruck quite as quickly, but eventually it will happen.

Dark-coloured glass can slow the rate at which beers become lightstruck, but only by a certain amount. Brown affords the best protection. Green glass, blue glass or clear glass do not stop UV-B. Some large commercial breweries use a hop extract called 'Tetra Hop', which is processed to exclude the chemical that changes in light; but to home-brewers this is not easily available.

The best possible protection for homebrew is to store beer in a dark place, away from direct sunlight or fluorescent lights. Unfortunately, once beer has become lightstruck there is no good way to remove the flavour. Open containers may dissipate the aroma in time, but it will never quite go away entirely.

🍂 *Dark-coloured glass can slow down the rate at which beer becomes lightstruck, but light will eventually get through even brown bottles, causing skunky beer, if they are left for too long in direct contact with UV-B rays.*

WHAT IS AN OFF-FLAVOUR?

A particular flavour that is present in some, or even all beer, becomes an off-flavour when it is overwhelming, or present in the wrong style. A common flavour in beer, isoamyl acetate — which tastes like banana-flavoured sweets — is appropriate in a hefeweissen, but doesn't belong in an oatmeal stout. There are many variables that contribute to off-flavours. Keeping them under control can be the difference between a good beer and a great one.

85 My beer smells/tastes like sewer or vomit

CAUSE

Butyric acid has been formed during the brewing process, either from rancid grain or an exceptionally long mash, such as a sour mash process.

SOLUTION

Butyric acid is one of the most unpleasant off-flavours that can be experienced in home-brewing. While it can sometimes be introduced to a brew when a brewer uses rancid or rotting grain, most home-brewers know better than to use grain that smells or looks gross.

Most often, butyric acid is formed when a brewer allows a mash to sit for a long time at a low temperature. In some cases, a brewery may be doing this on purpose to create a sour mash (also known as a sour kettle). Some brewers mash overnight to save time and allow for maximum possible sugar conversion in their mash.

When doing so, however, all precaution should be taken to reduce the oxygen in the environment. In the presence of oxygen, lactobacillus can create butyric acid, which can survive boiling and fermentation and exist in the final product. Of course, lactobacillus is not the only bacteria to live on barley. In some cases, clostridium may also be present in grain, which, when in an anaerobic environment, can create butyric acid. However, if the pH drops quickly enough, due to the presence of lactobacillus, clostridium will not survive.

To best avoid butyric acid in your brew, finish your mashes in a reasonable amount of time (under a day, at a consistent temperature). If souring, create an anaerobic environment that favours lactobacillus growth (32°C–37°C/ 90°F–100°F, with a blanket of CO_2) and monitor the pH level.

Be sure to start with fresh ingredients in your mash. Spend time looking at and tasting them. Sub-par or spoiled ingredients make sub-par beer.

86 My beer smells/tastes metallic

CAUSE

High-iron content in water, or brewing with a pot that might easily corrode in an acidic environment, may lead to beers with a metallic character.

SOLUTION

It is important to distinguish 'metallic', as an off-flavour, from 'bitterness'. Humans often perceive bitterness, or even high amounts of CO_2, as slightly metallic. In the case of beer, a metallic off-flavour will usually manifest itself in an tin-foil, copper-penny or blood-like flavour.

If your water is provided by a water company, ask them for a water report and then get the water in your home tested to compare. If you are using well water, you will need to send water out to a lab for testing. If your home water is high in iron or other metallic ions, this may translate to a metallic flavour in your beer, but it may also indicate a problem with your home plumbing that should be addressed.

In some rare instances, metal pots need to be passivated or seasoned before using for brewing, particularly if you are using a pot that is not made from stainless steel. Be careful of using anything that cannot handle an acidic environment, or which might corrode in contact with an oxidative cleaner – unless properly cared for, these can release metal ions into your beer.

If you are scrubbing the inside of your pot for any reason, avoid steel wool which can leave tiny scratches in the side of even stainless steel kettles; these can allow stainless steel to pit and corrode further, damaging your kettle and possibly affecting the flavour of your beer.

Avoid using a kettle made of anything other than stainless steel. Aluminium pots are inexpensive but long exposure to heat and the acidity of wort can break the aluminium down and create off-flavours.

87 My beer smells/tastes boozy

CAUSE

High fermentation temperatures, or over-oxygenation, has caused the yeast to create more fusel alcohols, which can lead to a boozy or 'hot' alcohol flavour.

SOLUTION

Sometimes a high-alcohol beer just tastes boozy because there's a lot of ethanol in it, but this should be rare.

Yeasts all have an ideal fermentation temperature range, but that doesn't mean that they can't ferment outside of that range. If yeasts ferment at a higher temperature, particularly right at the beginning of their growth cycle, they can sometimes create a high portion of fusel alcohols. These alcohols often taste like almonds in good cases, or in bad cases like solvent or airplane glue. In all cases they can read as 'hot' or 'boozy'.

If using an oxygenation stone, be careful not to overdo it. Dissolved oxygen content should be right around 8 ppm, which most home-brewers can achieve in a minute or so of oxygenation with an oxygen stone. Similarly, shaking a demijohn after filling it to get oxygen into solution will, in many cases, result in a dissolved oxygen of close to 8 ppm. The only way to be sure is to buy an oxygen meter, which can be an expensive proposition. In general, it is difficult to over oxygenate.

Fortunately, fusel alcohols, if allowed to age, will eventually break down into esters. However, that process can take months or even years. If you've made a particularly big beer and you're happy to put it away for ageing, you could be in for a treat, but in other beer styles you may just have to settle for the hot alcohol characters.

🌺 *Beer can be fermented in any size of vessel. Small fermenters allow a brewer to split larger batches, and experiment with different yeasts or other ingredients.*

88 My beer smells/tastes like vinegar

CAUSE

The beer has been infected with acetobacter, which creates acetic acid, commonly known as vinegar.

SOLUTION

Once infected with acetobacter, there is no way to salvage a beer. Unfortunately, from this point on, you have a malt vinegar. Acetobacter is an aerobic bacterium that, among other things, metabolises ethanol to create acetic acid. Therefore, if you have encountered an infection of acetobacter, it is safe to assume that there was an infection at a step that involved oxygen and ethanol in combination.

Yeast creates CO_2. Since CO_2 is heavier than oxygen, fermentation occurs in an aerobic environment. However, once fermentation is finished, along every step of the way the possibility of infections exists. Check that your bottling bucket is clean and does not smell vinegary or sour. You may need to check or replace your bottling hoses or bottling tip. Be sure that all bottles and caps have been sufficiently cleaned and sanitised prior to using.

Some beer styles, such as Flanders Red or Flanders Oud Bruin, feature yeasts or mixed bacterial cultures that can create a tiny bit of acetic acid, which blends with the residual sweetness of the beer, leaving it with a pleasant, sour finish. However, are should be taken when using yeasts in an otherwise 'clean' yeast brewery, .

My beer smells/tastes like soy sauce

CAUSE
Yeast autolysis has occurred because the beer has been stored while on yeast in a warm environment, and for a long period.

SOLUTION
Yeast autolysis occurs when yeast has run out of resources in its environment, and begins to consume its own internal resources, usually leading to cell death and the popping (lysing) of the cell. This releases the interior contents of the cell into the beer, which leaves a brothy or 'soy sauce'-like character.

The best way to avoid this is to move beer off its original yeast if it's going to be stored in a warm place for a long period. If kept cold, yeast will go dormant instead of continuing to scrounge for resources. If left too long in these conditions, it will die but will be unlikely to lyse. In general, it is best not to store finished beer in a warm environment, for many reasons in addition to this one. If it must be done, do what you can to minimise the amount of yeast that is left in the solution before storage.

Once this flavour is in your beer, there is no good way to get rid of it. Ageing may soften it, or blend it into other flavours, but if there is still yeast in solution it is equally likely that the flavour will intensify. In some high-gravity or barrel-aged beers, this flavour is considered a desirable characteristic at very low levels, as it adds a depth of flavour, but in high concentrations it can be a distraction to the overall product.

90 My beer smells/tastes like sulphur

CAUSE

Fermentation was not vigorous enough to remove sulphur from the liquid; or lager was not given enough maturation time before packaging.

SOLUTION

All yeasts create some amount of hydrogen sulphide – the sulphur compound that smells like rotten eggs or burnt matchsticks. However, ales generally ferment vigorously enough to quickly off-gas those compounds at the beginning of fermentation.

If you are not experiencing a very robust fermentation when using ale yeast, it is likely that you have underpitched it, or that the yeast is starved of nutrients. Using yeast nutrients – which are normally a mixture of dead yeast cells and zinc, and available from your local homebrew supplier – can help ensure that your fermentations are never in want of nutrients.

Lagers are not so lucky. They take a long time to ferment at cool temperatures, and often do not release gas at quite the same rate as their ale-yeast cousins do, although lager makes up for this with its long conditioning and maturation time. Part of that maturation is the oxidation and degradation of hydrogen sulphide. If your finished lager tastes overly sulphuric, it is likely that you just bottled too soon.

In the case of both ales and lagers, sulphur is something that will age out of beer relatively quickly, so long as it is not in the finished package. Once bottled, even if hydrogen sulphide does oxidise, the sulphur gas has nowhere to go. Some of it will likely off-gas once the beer is poured into a bottle, but some amount of it will always be there.

❧ *If a finished beer has been left with a sulphur-like quality, the fermentation may not have been vigorous enough. Beer should be free from sulphur before it is bottled to avoid an unpleasant 'rotten-egg' smell in the final product.*

91 My beer smells/tastes like wet newspaper/cardboard

CAUSE

Beer has aged, going through an oxidation process that forms a chemical called trans-2-NONENAL (T2N), which tastes like wet newspaper or cardboard.

SOLUTION

This is one of the main reasons why beer is best consumed fresh. Over time, as beer sits in a bottle, it slowly oxidises, creating the chemical T2N. This will eventually happen to all beer. At times, it will also occur alongside other, more pleasant ageing characteristics, but there is ultimately no way to completely avoid oxidation. It is, however, possible to slow down the formation of T2N.

Never store beer on its side. Even the best-sealed oxygen-scavenging bottle caps are porous, and will allow a small amount of oxygen to get into the bottle. If the bottle is sitting upright, a small CO_2 blanket should be present inside the bottle that will offer protection against oxygenation and oxidation. If the bottle is on its side, beer is in direct contact with the cap, which means that oxygen can penetrate the beer directly.

Keep beer in as cool a place as possible. It's often impractical to keep an entire batch of homebrew refrigerated unless it's in a keg, but every effort should be taken to keep beer as cool as possible, as cold temperature slows reactions.

Luckily, few home-brewers can filter beer completely clean, which means that there is almost always some amount of yeast in solution in the bottle. Yeast in the bottle can help scavenge oxygen and lower dissolved oxygen content at time of bottling, meaning a longer shelf life for your homebrew.

🍂 *Beer should be stored upright to avoid oxidisation and the formation of trans-2-NONENAL.*

Q10 OR THE ARRHENIUS EQUATION

The Arrhenius equation states that for every 10°C (50°F) rise in temperature, reactions happen twice as quickly. Thus, beer stored at 20°C (68°F) ages roughly four times faster than beer kept at 3°C (38°F). That means a week at room temperature ages a beer the equivalent of a month in a refrigerator. Make sure you keep your beer (and all beer) cold!

92 My beer tastes burnt

CAUSE

Wort was scorched during boil. When adding sugar or extract to a kettle, sugar will, at times, settle to the bottom of the kettle before dissolving into solution, burning on the bottom of the kettle.

SOLUTION

Heat management can be tricky with home-brewing, particularly when brewing on an electric hob – in which coils are in direct contact with the kettle – or when brewing with a high-BTU outdoor burner. To help diffuse the heat across the bottom of the kettle, consider using a diffusion plate – a metal disc with many bumpy points on it that sits on top of an electric coil or outdoor burner.

Consider heating more slowly. Most burners allow you to regulate the amount of heat being generated, and it may be that bringing wort to a boil more slowly avoids scorching. However, once the wort has reached boiling point, be sure to maintain a vigorous boil, to remove as many off-flavours as possible.

When your kettle is on the heat, even if the wort is not yet at boiling point, carefully add any sugar or extract, stirring slowly so that the sugar has time to dissolve into solution before hitting the bottom of the kettle. This will avoid any potential boil-overs. Many syrups or heavy sugars will burn on contact with the bottom of the kettle, and once the flavour of burnt malt has been introduced into your beer, it is there permanently. If possible, draw from your kettle some hot wort or hot water to add to the sugar or extract, dissolving it before adding, or preheat any syrups to make them more readily able to dissolve into hot liquid. They'll also be easier to pour this way.

All beer needs to be boiled but sometimes direct heat from gas flames are so hot that it's easy to scorch the wort.

93 My hoppy beer is too bitter and lacks flavour

CAUSE

The beer recipe is scaled towards early bittering additions, rather than late-flavour and aroma additions.

SOLUTION

As a general rule of thumb, hops added at the beginning of the boil will result in a more bitter beer, while hops added at the end of the boil will result in a more aromatic or flavourful beer.

When choosing hops, consult a vendor's website, which can provide the expected ranges of hop oils. Look out for total oil content, B-pinene (piney flavours), myrcene (piney/resinous), linalool (floral/citrus), caryophyllene (woody), farnesene (floral), humulene (woody), 3-mercaptohexanol (tropical fruit), 3-mercaptohexyl acetate (tropical fruit), 4-mercapto-4-methyl-pentan-2-one (berries, fruit) and geraniol (rosewater). There are, of course, many more oils that provide flavours in hops, but these are some of the most commonly listed and noted oils.

Make sure you match carefully when substituting hops with different alpha-acid contents (see Problem 22), particularly when substituting high-alpha hops for late additions in the boil. Bear in mind that while the beer is over 79°C (175°F), isomerisation is taking place, which includes the entire whirlpool process of the beer. Consider adding late-addition hops at the end of the boil, or even in whirlpool, particularly if you experience very vigorous boils. You may be driving off the extra flavours and aromas you're seeking.

Finally, consider dry hopping, once or a few times (see Problem 52) for added aroma and flavour. Using a blend of hops with a high oil content can add a considerable amount of depth to the flavour and aroma of any beer, not just one that's supposed to be hoppy.

Large hop additions are very popular among home-brewers, but hops should be added with care to produce the desired balance of flavour and bitterness.

94 My beer is hazy when cold

CAUSE

Beer is experiencing chill haze, a fine matrix of proteins that stick together when cold, but dissolve into liquid as the beer warms.

SOLUTION

Fortunately, chill haze is only a problem if you prize exceptionally crystal-clear beer. It does not affect flavour or body; it is merely a haze in solution.

Chill haze forms when there has not been enough of a 'cold break' at the end of the boil. 'Cold break' refers to the precipitation of proteins that happens as wort cools from the boil. In that process, proteins and carbohydrates stick together and fall out of solution into trub. If the beer is not cooled quickly enough, or if a large amount of trub is introduced into fermentation, some of those proteins will dissolve back into the liquid and cause this light haze when the beer is cold.

To avoid chill haze, consider adding finings to the last few minutes of the boil. Home-brewers most commonly use Irish moss, which is a seaweed (also known as *Chondrus crispus* or carrageenan moss). It works by creating a negative charge within wort, which positively charged proteins stick to. The larger the mass of protein and seaweed, the more that gravity acts to drop them out of solution, at which point a careful home-brewer can transfer wort away while leaving all the protein behind.

Chill haze is a cosmetic problem. Odourless and tasteless, it will not change the flavour or shelf stability of your beer. As the beer ages, it is less likely to experience chill haze, as the proteins will eventually drop out of solution.

🌸 *When proteins and yeast proteins don't coagulate, and float on top of fermentation, they can lead to a hazy beer.*

95 My dark beer tastes ashy

CAUSE

There is too much roasted malt in the recipe, or the pH of the mash or boil was too low, heightening the tannin/roasted character of the roasted barley husks.

SOLUTION

In most cases, if a beer comes out tasting ashy, it is because the recipe called for a high proportion of roasted malt, particularly very dark, roasted malts like Black Patent or Black Malt.

Speciality malts such as dark roast malts should be used in very small quantities. As they do not add fermentable sugar to the wort, high proportions of roasted malt can add dextrins, creating a more robust body, and sometimes contribute residual sugars for a sweeter beer. However, they can also contribute a lot of tannic content from the roasted husk of the barley. Since the husk on roasted barley is more brittle, it crushes more easily, which means that it has greater surface area contact, and can have a much higher extraction rate. Most well-built recipes contain a very small percentage of dark roasted speciality malts with some recipes calling for as little as 1 per cent of the total grist.

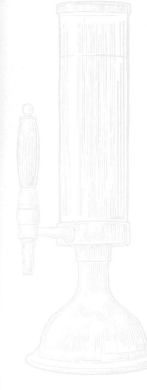

Some home-brewers add dark grains to the mash at the very end to extract as little of the tannin bitterness as possible, or rely on huskless roasted grains such as roasted wheat, roasted rye or de-husked barley for dark character without tannin bitterness.

A good mash should have a pH content of around 5.2–5.4. Below that, you run the risk of extracting additional tannin bitterness from dark roasted grains. If you are unintentionally achieving a pH below 5, check the starting pH of your brew water, and consider adjusting it using calcium carbonate or sodium bicarbonate.

🌺 *Dark grains are beautiful, but a little goes a long way. The burnt husk of barley can contribute tannins and harsh bitterness to beer.*

96 My beer is more/less bitter than I expected.

CAUSE

The alpha acid on the hops does not match that on the recipe, brewing water has not been adjusted to match the stated outcome in the recipe.

SOLUTION

When purchasing hops, on most occasions, the important factor that brewers pay attention to is variety. If a recipe calls for Cascade hops, a brewer buys Cascade hops. They may not pay attention to the alpha-acid content listed on each package. Cascade (7.5 per cent AA) will produce considerably more bitterness than Cascade (4.5 per cent AA), yet both exist.

The degree of alpha-acid content depends upon what field the hops were grown in, in what region of the world, how much rainfall or sunlight the hops received while they were growing, and even what time they were harvested.

It is up to the brewery to adjust the calculations in their recipe to match bitterness, based on expected IBU (see Problem 22). However, in some cases, when following a recipe, bitterness may come down to water hardness. Water with a high mineral content normally favours a harsher bitterness in hoppy beers, while soft water subdues bitterness and helps highlight malt characteristics. To achieve a certain bitterness profile through a combination of both hops and minerals, it is often easiest to start with purified water and add brewing salts to create the target hardness level.

There are multiple online calculators and pieces of software to help aid in this process, as well as to match your brewing water with that of the water in the towns of classic brewing styles, such as Dublin, Munich or Burton-on-Trent.

❧ When it comes to hops, there are a great variety of options available to home-brewers. As a natural product, there are also subtle variables that affect their alpha-acid content. Whole-cone hops, pictured here, are beautiful and delicate.

97 My beer is too sweet/tastes like wort

CAUSE

Fermentation did not start because of a problem with yeast viability or vitality, did not finish completely due to pitch rate problems or temperature stability, or hops were not added.

SOLUTION

In the best-case scenario of fermentation issues, the beer ends up sweet and insufficiently attenuated, but not infected with anything, leaving you in a good place to recover from: just pitch more yeast and allow the fermentation to finish out naturally.

The root cause of under-attenuated or under-fermented beer is often yeast health. Be sure that the yeast that you pitched was fresh and not expired and that enough has been pitched. In general, when using shop-bought yeast pitches, roughly one package per 20 litres (5 gallons) per 10 degrees plato is usually a great starting point (see Problem 21). When in doubt, err on the slightly high side, or use a yeast starter (see Problem 20) to ensure robust fermentation.

Be sure that the fermenter is being kept in a room that doesn't get too cold. Even though yeast generates a fair amount of its own heat while fermenting, a cold ambient environment can stall fermentation and cause the yeast to go dormant. Returning the yeast to a warm temperature should kick-start fermentation, though with very flocculent strains the yeast may need to be roused to get back into solution.

Finally, be sure that hops were added at the time specified on the recipe. Even beers with no discernible hop flavour or aroma need hops to help curb sweetness. Hops help beer taste like beer. A home-brewer who does not have a palate for hops may be surprised to find out just how much they rely on hops for the character of their beer.

🌿 With most problems relating to off-flavours in the final beer, a home-brewer often needs to retrace their steps to ascertain what might have gone wrong. What temperature was the fermenter kept at? When were the hops added to the liquid? Keeping a notebook can help with remembering these details.

My beer doesn't have good head retention

CAUSE

Beer has been mashed at too low a temperature, beer lacks enough hop presence to contribute polyphenols for good head retention, or there is residual soap residue in fermentation vessels.

SOLUTION

It might not be intuitive to think so, but good head retention is usually the product of careful mashing. A mash on the warm side of the spectrum, between 66°C (150°F) and 68°C (154°F), or even a warmer mash at around 69°C (156°F) will promote dextrin formation, which can lead to a finished beer with a fuller body and better head retention. Low head retention is often due to a beer mashed too cool, which promotes the formation of fermentable sugars but not of body-building dextrins.

Hops also contribute polyphenols to wort, which can help with both clarity and head retention. Beers with only small hop additions often suffer from poor head retention. If your recipe can handle it (or you really like hops), consider adding in some more hops to help build head retention.

Consider adding high-dextrin malts, such as Cara-Pils, Cara-Helles, Dextrin Malt or even very light Caramel Malts, to help build body and head retention. You may also use high-protein adjuncts such as rye, wheat, oats or even flaked barley to help build a protein matrix that can support a good head.

Finally, be sure that all soap residue has been cleaned off your equipment before brewing. Residual surfactants can cause surface tension in beer to break, which all but eliminates head retention.

🍃 *All home-brewers aspire to produce the perfect-looking beer, with good head retention. Stemmed glassware (pictured) is a popular choice for beer because you can hold the stem, rather than the bell of the glass, which slows the rate at which the beer warms in the glass.*

THE IMPORTANCE OF GOOD HEAD

A head on a beer is a great source of aroma and, thus, a great source of flavour. The bubbles at the top of a beer are little pockets of aromatic gases and hop oils, waiting to be released into your olfactory nerve when you sniff. Since aroma comprises most of what you're tasting, it could be argued that a beer with a good head tastes better.

99 The body of my beer is too thin

CAUSE

Beer has been mashed at too low a temperature or a large amount of fermentable adjunct has been added to the beer without a corresponding rise in protein and dextrin.

SOLUTION

Like head retention, a good body comes from a profile and recipe with a balanced mash temperature. Temperatures in the higher range of mash temperatures closer to 68°C (154°F) will often promote a fuller-bodied beer. The addition of high dextrin malts, or malts with a higher number of complex, unfermentable sugars, can also contribute to a full body (see Problem 98).

A thin body can also be the result of adding in highly fermentable adjuncts such as maize or rice, or even table sugar or Belgian Candi sugar. These additions promote fermentation and alcohol production, but do not add to the complex sugar or protein structure of the beer, which can mean a thin, somewhat cidery body.

When brewing beers with highly fermentable adjunct additions, consider mashing warmer than you normally would, or adding in high-protein grains, such as wheat, oats or flaked barley, to help create a balanced body.

Thin bodies are most often an issue when making low-alcohol or 'session' beers. To create a good, rich body in low-alcohol beers, assume a decrease in mash efficiency of about 15 per cent and mash much warmer, at around 70°C (158°F). A higher proportion of barley will give you the same amount of fermentable sugars, but with the added bonus of more dextrin to help create a more robust and fuller body. Bear in mind that you should only increase base malt in this case, and that the beer may turn out slightly darker than expected.

❧ *Tulip glasses (pictured) are excellent for beer. They have a large bell to hold the beer from without warming it, and the flare at the top of the glass will support a good head while releasing aroma.*

SESSION BEERS

Session beers are lower-alcohol beers of between 3 to 4 per cent ABV, meant to be more suited to drinking in volume during a session. They are most usually Bitters or Milds but can be in other styles. With styles that are generally stronger, the word 'session' may be used to indicate that a particular beer is a lower-alcohol version, such as a 'Session IPA'.

100 I'd like impartial feedback about my beer

CAUSE

It's great to share your beer with friends and hear great reviews, but impartial feedback from experienced tasters is often the key to making better beer.

SOLUTION

Find your local homebrew club. Most clubs meet on a regular basis, and give you a chance to sit down and share your brew with other people with the same hobby. Most homebrew club meetings will feature a wide range of people, from experienced homebrew veterans who have been brewing for decades, to new beginners who are on their first kit.

In all, it is an excellent way to find new tips and tricks, discover new styles and ideas, and share some great beer with like-minded hobbyists. They are often the best first people to give you critical feedback about your beers along with ideas on how to make it better.

You may want to consider entering a homebrew competition. These competitions are often organised by style and will require the brewer to meet guidelines. Each country and region of the world has its own set style of beer guidelines that can be found either on the Internet or with the help of your local homebrew club.

Homebrew competitions give you the chance of getting the most impartial blind feedback on your beer. It will be tasted alongside others of the same style in a blind panel by a trained judge. While they can't comment on your process, they can give feedback on the flavour and composition of your finished beer, and how well it fits style guidelines. These can be helpful learning experiences and, who knows, you just might win.

TRAIN TO BE A BEER JUDGE

There are various training courses and workshops that can be attended to learn how to judge beer. Many lead to an exam through which a certificate can be obtained. The American-run Beer Judge Certification Program (www.bjcp.org) is one of the most well-respected and well-organised beer judging programmes internationally. They hold workshops and exams around the world, including in the UK. Within the UK, recognised training is also available through the Beer Academy (www.beeracademy.co.uk) who, among other courses, run a one-day course entitled, 'How to Judge Beer.'

It's not snobby or silly to smell a beer; you're actually getting a great preview of what it's going to taste like.

GLOSSARY

ACETALDEHYDE
A common off-flavour that has a green-apple or mown grass-like flavour and aroma.

ALCOHOL BY VOLUME
A measure of how much alcohol is contained in a standard volume of an alcoholic beverage.

ALCOHOL BY WEIGHT
A measure of how much alcohol is in an alcoholic beverage expressed as a percentage of the total beverage weight.

ALE
Refers to beer made with ale yeast (saccharomyces cerevisiae), which ferments better in warmer temperatures, creating a full-bodied, often fruity, flavour profile.

ATTENUATION
The reduction of specific gravity in wort caused through fermentation when yeast consumes wort sugars and converts them into alcohol and CO_2.

BOTTLE-CONDITIONED BEER:
Beer that is bottled with a viable yeast population in suspension and undergoes an additional fermentation in the bottle, giving natural carbonation and, as such, meets CAMRA's definition of real ale. Recently CAMRA has recognised that some canned beer also meets this criterion.

BRETTANOMYCES
A genus of single-celled yeast organisms that ferment sugar, and can produce desirable or undesirable characteristics.

CASK-CONDITIONED BEER/CASK ALE:
Unfiltered and unpasteurised beer that continues to ferment and condition within a cask from which it is served without the use of additional CO_2 or nitrogen pressure. In most pubs such beers are served through a mechanical hand-pump. Cask-conditioned beer is real ale.

DEMI-JOHN
A large bottle – glass, plastic, or earthenware – used for fermentation of wort.

DEGREES PLATO
A measure of the density of liquid, invented by Karl Balling in 1843, and refined later by Fritz Plato.

DIACETYL
A common off-flavour that has a butter or butterscotch-like flavour and aroma.

DIMETHYL SULPHIDE (DMS)
A common off-flavour that has the aroma and flavour of cooked vegetables.

FUSEL ALCOHOL
A family of high molecular weight alcohols caused by excessively high fermentation temperatures, imparting harsh characteristics, resembling paint thinner.

GRIST
Part of the beer recipe that is malt, particularly after it's crushed; also known as the 'grain bill'.

HOPS
The flowering and fruiting body of the plant *Humulus lupulus*, used to impart bitterness and flavour to beer.

IBU
International Bitterness Unit; One part-per-million of isomerised alpha acid, or 1 mg/litre.

KEG
A cylindrical vessel that holds carbonated beer under pressure. Homebrew kegs are generally 5 gallons in volume.

KRAUSEN
The head of foam which

appears on the surface of wort during fermentation.

LACTOBACILLUS
Bacteria that converts unfermented sugars found in beer into lactic acid. Sometimes thought to spoil beer, but can be intentionally added to finished beer in order to add desirable sourness.

LAGER
German term meaning 'to store'; in brewing, refers to beer made with lager yeast (saccharomyces pastorianus), which ferments better in cooler temperatures, creating a more crisp, less fruity flavour profile.

LAUTERING
The process of removing sugary liquid from the mash vessel.

LIQUOR
Refers to water used in the brewing process before grain has been infused into it.

MAILLARD REACTION
The result of applying heat to brown food and drink, creating flavour and changing the colour.

MALT
Barley forced through a false germination process, then kilned, creating a reserve of starch and enzymes that will transform into sugar during the mashing process.

MALT EXTRACT
Powdered or liquid malt made to take the place of mashing/grain in a homebrew setting.

MASH
The mixture of grain and hot water that forms the wort.

MASH TUN
The vessel in which the grain is soaked in water and heated, converting starch to sugar and extracting sugars, colours and flavours for brewing.

PEDIOCOCCUS
Bacteria usually considered contaminants of beer, although their presence is desirable for some beer styles.

SACCHAROMYCES
The genus of single-celled yeasts that ferment sugar and are commonly used in brewing.

SPARGING
The action of rinsing hot water over the top of the mash filter bed during lauter, which assists in removing sugar from the grain.

SPECIFIC GRAVITY
A measure of the density of liquid expressed in a ratio of the density of a substance to the density of a reference (eg. wort-to-water).

TRANS-2-NONENOL (T2N)
A common off-flavour that has the flavour of wet newspaper or cardboard.

TRUB
Residual proteins, hop oils and tannins produced during the boiling and cooling stages of brewing.

VORLAUF
The process of recirculating liquid from the bottom of the mash vessel to the top of the mash vessel, in order to set the grain into a filter bed.

WORT
Unfermented beer; specifically the liquid that will become beer, before yeast has been added to it.

YEAST
A unicellular fungus that metabolises sugar and secretes ethanol, CO_2 and hundreds of flavour compounds that make beer both alcoholic and delicious.

A-Z OF BEER VARIETIES

BARLEY WINE

Dates from the 18th and 19th centuries when England was often at war with France and it was the duty of patriots, usually from the upper classes, to drink ale rather than French claret. Barley wine had to be strong – often between 10 and 12 per cent – and was stored for up to 18 months or two years. Expect massive sweet malt and ripe fruit of the pear drop, mandarin orange and lemon type, with chocolate and coffee if darker malts are used. Hop rates are generous and produce bitterness and peppery, grassy and floral notes.

BITTER

Originally produced from the turn of the 19th and 20th centuries as a new type of running beer that could be served after a few days of conditioning in pub cellars (brewers built large estates of 'tied' pubs and moved away from beers stored for months or years). Bitter developed from Pale Ale but was usually copper coloured or deep bronze due to the use of slightly darker malts, such as crystal, that gave the beer fullness of palate. Best is a stronger version of Bitter but there is considerable crossover. Bitter falls

into the 3.4 to 3.9 per cent alcohol band while Best Bitter is upwards of 4 per cent (though a number of brewers call their ordinary Bitter 'best') and some brewers make Strong Bitter of 5 per cent or more. Ordinary bitter has a spicy, peppery and grassy hop character, a powerful bitterness, with tangy fruit and juicy/nutty malt. Best and Strong Bitters have a dominant malt and fruit character but also a hop aroma and bitterness, often achieved by 'late hopping' during the boil or by adding hops to casks as they leave the brewery.

BURTON ALE

A popular style from the 18th and 19th centuries that originates from Burton-on-Trent, UK, where at one time six different versions were produced that ranged from 6 to 11.5 per cent alcohol. In the 20th century, Burton was overtaken in popularity by Pale Ale and Bitter but was revived in the late 1970s with the launch of Ind Coope Draught Burton Ale. When Allied Breweries broke up, the beer was owned by Carlsberg, who stopped production in 2015. The style has now been recreated in its hometown by Burton Bridge Brewery. Other versions of

the style exist under different names: Young's Winter Warmer was originally called Burton. Bass No 1, brewed occasionally, is called a barley wine but is in fact the last remaining version of a Bass Burton Ale. Look for a bright amber colour, a rich malt and fruit character underscored by a solid resinous and cedar wood hop note.

FRUIT/SPECIALITY BEERS

The popularity in Britain of Belgian fruit beers has led to many UK brewers adding fruit to their beer in a bid to seek out a wider audience. Honey, herbs, heather, spice and even spirits also feature in a number of speciality beers, while beers matured in Bourbon, whisky and Cognac casks have become a major development in both this country and the US. Fruit and honey beers are not sweet; the ingredients add new dimensions to the brewing process and are highly fermentable. Beers that use the likes of cherries or raspberries are dry and quenching rather than cloying.

GOLDEN ALE

Now so popular with both drinkers and brewers that the style has its own category in the annual Champion Beer of Britain competition. Since the early 1980s this beer was produced to wean younger drinkers from mass-produced lager to the pleasures of cask ale. It is paler than a pale ale, often brewed with lager malt or specially produced

low-colour ale malt. As a result, hops are allowed to give full expression, balancing sappy malt with luscious fruit, floral, herbal, spicy and resinous notes. Golden ales often feature imported hops from North America, the Czech Republic, Germany, Slovenia and New Zealand, rather than the traditional English hops used in pale ale, giving them radically different notes. They are often served colder than draught Bitter.

INDIA PALE ALE (IPA)

India Pale Ale changed the face of brewing in the 19th century. The new technologies of the Industrial Revolution enabled brewers to use pale malts to design beers that were pale bronze in colour. The first 'India Ales' were brewed in London and were probably based on October Beers that were matured for many months and were ideally suited to a long sea journey to India. But London was soon eclipsed by Burton-on-Trent, with its spring waters rich in minerals that brought out the fullest flavours of malt and hops. The 19th-century IPAs were high in both alcohol and hops to keep them in good condition during the journey to the colonies. IPA's life span was brief, being driven out of Africa

and India by German lager beer. But the style has made a big comeback in recent years and is now made in abundance throughout the world. Look for a big peppery hop aroma and palate balanced by juicy malt and tart citrus fruit.

MILD

Usually dark brown in colour, due to the use of well-roasted malts or roasted barley (though there are paler versions) with a rich malty aroma and flavour, and hints of dark fruit, chocolate, coffee and caramel. There is also a gentle underpinning of hop bitterness. Mild was developed in the 18th and 19th centuries and drunk primarily by industrial and agricultural workers who needed refreshment but wanted a slightly sweeter and less aggressively hopped beer than Porter. Early Milds were much stronger than modern versions, which tend to fall between 3 and 3.5 per cent, though the stronger style is again becoming popular.

OLD ALE

A style from the 18th century, stored for many months or even years in wooden vessels where the beer picked up some lactic sourness from wild yeasts and tannins in the wood. Because of the sour taste it was dubbed 'stale', and the beer was one of the components of the early Porters. In recent years, Old Ale has made a return to popularity. It does not have to be especially strong and can be no more than 4 per cent alcohol. Neither does it have to be dark: Old Ale can be pale and bursting with lush malt, tart fruit and spicy hops. Darker versions have a more profound malt character, with powerful hints of roasted grain, dark fruit, polished leather and fresh tobacco. This style has a lengthy period of maturation, often in bottle rather than cask.

PALE ALE

According to a legend in the 19th century, when a sailing ship bound for India with a cargo of IPA foundered off the coast at Liverpool, the casks were brought ashore and news of both the colour and taste of Pale Ale spread throughout the country. IPAs were brewed for the domestic market as a result but the Burton brewers were keen to produce versions with lower alcohol and hop rates that didn't require months to mature. The spread of the railway system allowed brewers in Burton to

move beer around the country at speed and Pale Ale was dubbed 'the beer of the railway age' as a result. The clamour for Pale Ale was so great that brewers from London, Liverpool and Manchester opened second breweries in Burton to make use of the mineral-rich water to make their own versions of the style. From the early 20th century, Bitter began to overtake Pale Ale in popularity and a result Pale Ale became mainly a bottled product. A true pale ale should be different to Bitter, similar in colour and style to IPA and brewed without the addition of coloured malts. It should have a spicy/resinous aroma

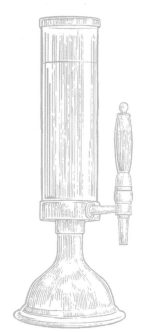

and a palate with biscuit malt and tart fruit from the hops. Many beers called Bitter today should properly be labelled Pale Ale.

PORTER/ STOUT

Jet-black in colour, with a dark and roasted grain character with burnt fruit, coffee, liquorice and molasses, this beer variety should have a deep bitterness to balance the richness of malt and fruit. Milk Stout, made by a few brewers, uses lactose or 'milk sugar' to give a creamy character. Porter was a London beer, the first in the world to be commercially brewed (in the early 18th century). Its name came from its popularity with London porters who needed calories to help sustain them in their hard manual labour. The origins of the beer are disputed but recent research suggests Porter, first called 'Entire', was blended in the brewery from pale, mild and aged or 'stale' beer. The strongest version of Porter was called Stout Porter, later shortened to just Stout. Porter and Stout were exported from London to the rest of the British Isles and, as a result, Arthur Guinness built his own Porter brewery in Dublin. During the First World War, when the British government

restricted the use of malt – heavily roasted versions in particular – in order to divert grain and energy to bread making and the arms industry, Guinness and other Irish brewers came to dominate the market. In recent years, Porter and Stout have returned to popularity in Britain, the United States and Australasia, with brewers digging into old recipe books to create genuine versions of the style.

SAISON

A Belgian beer style now finding favour in Britain and other countries. It originates in Wallonia, the French-speaking region of Belgium, and was a seasonal beer brewed by farmers to refresh their labourers during the busy harvest period. In sharp contrast to lambic, saison should have a rich malty/fruity palate balanced by earthy, spicy and peppery hops. Some saisons are made with the addition of botanicals such as ginger, black pepper and aniseed.

SCOTTISH BEERS

Tend to be darker and maltier than beers south of the border, the reflection of a colder climate where beer needs to be nourishing. It's an urban myth, though, that Scottish beers are less heavily hopped than English ones. The classic traditional styles are Light, Heavy and Export, which are not dissimilar to Mild, Bitter and IPA. They are also known as 60, 70 and 80 Shilling ales from a 19th century system of invoicing beers according to strength. A 'Wee Heavy' or 90 Shilling Ale, now rare, is the Scottish equivalent of barley wine. Many of the newer brewers in Scotland are producing beers lighter in colour and with pronounced hop character.

SOUR BEER

Known simply as 'Sours' and produced in both the UK and US by brewers fascinated by Belgian lambic beer, made by 'wild' or spontaneous fermentation. Lambic beer is left open to the atmosphere to allow wild yeasts to attack the sugars in wort and begin the fermentation process. Following the first fermentation, true lambics are stored in wooden casks for a year or more. The main wild yeast, Brettanomyces, is used by modern brewers to inoculate their worts and gain the required sour or acidic character.

WHEAT BEER

A style closely associated with Bavaria and Belgium, its popularity in Britain has encouraged many brewers to add wheat beers to their portfolios. All 'wheat beers' are a blend of malted barley as well as wheat, as the latter is difficult to brew with and the addition of barley acts as a natural filter during the mashing stage. But wheat, if used with special yeast cultures developed for brewing the style, gives distinctive aromas and flavours, such as clove, banana and bubble-gum, that make it a complex and refreshing beer. The Belgian version of wheat beer often has the addition of herbs and spices, such as milled coriander seeds and orange peel – a habit that dates to medieval times.

INDEX

FURTHER RESOURCES

BOOKS

Daniels, Ray. *Designing Great Beers: The Ultimate Guide to Brewing Classic Beer Styles.* Boulder, CO: Brewer's Publications, 1996.

Papazian, Charlie. *The Complete Joy of Homebrewing* (3rd ed.). New York, NY: Harper Collins, 2003.

Mosher, Randy. *Radical Brews.* Boulder, CO: Brewer's Publications, 2010.

Palmer, John. *How to Brew.* Boulder, CO: Brewer's Publications, 2006.

Wheeler, Graham. *Brew Your Own British Real Ale* (illustrated 3rd ed.). St Albans, Hertfordshire: CAMRA Books, 2014.

ONLINE

A good place to improve your knowledge and see what others get up to is a home-brewing forum. A good and well-attended forum is: www.jimsbeerkit.co.uk

The major British home-brewers' organisation is the Craft Brewing Association. The CBA has a number of regional sub-groups through which members meet up It also issues a quarterly printed magazine. www.craftbrewing.org.uk

Brew Your Own magazine's website is full of great advice and tools: http://byo.com

Brewer's Friend has good advice and is also good for recipes: http://www.brewersfriend.com

ABOUT THE CAMPAIGN FOR REAL ALE

CAMRA, the Campaign for Real Ale, is an independent, not-for-profit, volunteer-led consumer group. We promote good quality real ale and pubs, as well as lobbying government to champion drinkers' rights and protect local pubs as centres of community life.

CAMRA has more than 185,000 members from all ages and backgrounds, brought together by a common belief in the issues that CAMRA supports and their love of good quality British beer.

From just £24 per year – that's less than a pint a month – you can join CAMRA and enjoy a range of member benefits. Visit our website **www.camra.org.uk/joinus** or call us on 01727 867 201.

CAMRA Books is the publishing arm of the Campaign for Real Ale. We publish books on beer, pubs, brewing and beer tourism. Our books are available to purchase direct from the CAMRA Shop or through your local bookshop. Visit **www.camra.org.uk/camrabooks** or call 01727 867 201.

IMAGE CREDITS

Cover © Ievgenii Meyer, Roger Siljander and Mariusz Szczygiel | Shutterstock; © standret | iStock

08 © LindaZ74 | Shutterstock

11 © Wavebreak Media | Alamy Stock Photo

13 © underworld | Shutterstock

15 © Kirill Z | Shutterstock

19 © Bdoss928 | Dreamstime.com

21 © Wavebreakmediamicro | Dreamstime.com

23 © Bmosh99 | Dreamstime.com

25 © Goss Images | Alamy Stock Photo

26 © Vaclav Mach | Shutterstock

29 © JVisentin | Shutterstock

31 © Infrequent_Flyer | Shutterstock

33 © MementoImage | iStock

35 © blickwinkel | Alamy Stock Photo

37 © MarisaPerez | iStock

39 © Gweedoe | iStock

41 © simonmcconico | iStock

43 © Infrequent_Flyer | Shutterstock

45 © jordieasy | iStock

47 © Mark Sawyer | Alamy Stock Photo

49 © Adam Barhan | Flickr Creative Commons

51 © Bmosh99 | Dreamstime.com

52 © BDoss928 | Shutterstock

55 © Qin Xie | Shutterstock

57 © Bmosh99 | Dreamstime.com

59 © RuslanDashinsky | iStock

61 © BDoss928 | Shutterstock

63 © matzaball | iStock

65 © Paul Narvaez | Flickr Creative Commons

67 © Roger Siljander | Shutterstock

69 © Wavebreak Media | Alamy Stock Photo

71 © Bdoss928 | Dreamstime.com

73 © KaraGrubis | iStock

77 © Joshua Rainey | Dreamstime.com

78 © Roger Siljander | Shutterstock

81 © Bdoss928 | Dreamstime.com

83 © Bmosh99 | Dreamstime.com

85 © Bmosh99 | Dreamstime.com

87 © Bdoss928 | Dreamstime.com

89 © Bmosh99 | Dreamstime.com

91 © Paul Narvaez | Flickr Creative Commons

93 © Benoit Daoust | Shutterstock

95 © Paul Narvaez | Flickr Creative Commons

96 © Bmosh99 | Dreamstime.com

99 © Jinx! | Flickr Creative Commons

103 © Shawn Hargreaves | Flickr Creative Commons

105 © Josh Delp | Flickr Creative Commons

107 © FoodCraftLab | Flickr Creative Commons

109 © Bdoss928 | Dreamstime.com

111 © RobinsonBecquart | iStock

113 © Shawn Hargreaves | Flickr Creative Commons

115 © rozbeh| Shutterstock

117 © AdamChandler86 | Flickr Creative Commons

119 © shellhawker | iStock

121 © Steve Holderfield | Shutterstock

122 © Steve Heap | Shutterstock

125 © naiserie | Flickr Creative Commons

127 © Graham Corney | Alamy Stock Photo

131 © Click_and_Photo | iStock

133 © AdamChandler86 | Flickr Creative Commons

135 © AdamChandler86 | Flickr Creative Commons

137 © Popartic | Shutterstock

139 © Dave Shea | Flickr Creative Commons

141 © Enolabrain | Dreamstime.com

143 © AdamChandler86 | Flickr Creative Commons

145 © Anders Peterson |Flickr Creative Commons

147 © Kyle Van Horn | Flickr Creative Commons

149 © Andrew Ressa |Flickr Creative Commons

150 © Bdoss928| Dreamstime.com

153 © matzaball | iStock

155 © Bmosh99 | Dreamstime.com

157 © BDoss928| Shutterstock

159 © BDoss928| Shutterstock

161 © Zimmytws | iStock

163 © ulcojan | Shutterstock

165 © Olaf Speier | Shutterstock

167 © JHK2303 | iStock

169 © Israel Patterson| Shutterstock

171 © Infrequent_Flyer | Shutterstock

173 © AdamChandler86 | Flickr Creative Commons

175 © BDoss928 | Shutterstock

176 © Wavebreakmedia | iStock

179 © Dave Shea | Flickr Creative Commons

181 © Benoit Daoust | Shutterstock

183 © Click_and_Photo | iStock

185 © jfarango | iStock

187 © BDoss928 | Shutterstock

189 © JVisentin| Istock

193 © rasilja | iStock

195 © RightOne | iStockJ

197 © Visentin | iStock

199 © Yuri_Arcurs | iStock

201 © Jeena Paradies | Flickr Creative Commons

203 © HQuality | Shutterstock

205 © Jacob A Bielanski | Shutterstock

207 © BDoss928 | Shutterstock

209 © Jarno Holappa | Shutterstock

211 © Bernt Rostad | Flickr Creative Commons

213 © Click and Photo | Shutterstock

All engravings © Shutterstock